T0135054

Springer Theses

Recognizing Outstanding Ph.D. Research

Aims and Scope

The series "Springer Theses" brings together a selection of the very best Ph.D. theses from around the world and across the physical sciences. Nominated and endorsed by two recognized specialists, each published volume has been selected for its scientific excellence and the high impact of its contents for the pertinent field of research. For greater accessibility to non-specialists, the published versions include an extended introduction, as well as a foreword by the student's supervisor explaining the special relevance of the work for the field. As a whole, the series will provide a valuable resource both for newcomers to the research fields described, and for other scientists seeking detailed background information on special questions. Finally, it provides an accredited documentation of the valuable contributions made by today's younger generation of scientists.

Theses are accepted into the series by invited nomination only and must fulfill all of the following criteria

- They must be written in good English.
- The topic should fall within the confines of Chemistry, Physics, Earth Sciences, Engineering and related interdisciplinary fields such as Materials, Nanoscience, Chemical Engineering, Complex Systems and Biophysics.
- The work reported in the thesis must represent a significant scientific advance.
- If the thesis includes previously published material, permission to reproduce this must be gained from the respective copyright holder.
- They must have been examined and passed during the 12 months prior to nomination.
- Each thesis should include a foreword by the supervisor outlining the significance of its content.
- The theses should have a clearly defined structure including an introduction accessible to scientists not expert in that particular field.

More information about this series at http://www.springer.com/series/8790

Alessia Benedetta Platania

Asymptotically Safe Gravity

From Spacetime Foliation to Cosmology

Doctoral Thesis accepted by
Università degli studi di Catania, Italy,
and Radboud University Nijmegen, The Netherlands

 Springer

Author
Dr. Alessia Benedetta Platania
Institut für Theoretische Physik
Universität Heidelberg
Heidelberg, Baden-Württemberg
Germany

Supervisors
Dr. Alfio Bonanno
Università degli studi di Catania
Catania, Italy

Dr. Frank Saueressig
Radboud University
Nijmegen, The Netherlands

ISSN 2190-5053   ISSN 2190-5061   (electronic)
Springer Theses
ISBN 978-3-030-07533-0   ISBN 978-3-319-98794-1   (eBook)
https://doi.org/10.1007/978-3-319-98794-1

This Springer imprint is published by the registered company Springer Nature Switzerland AG
The registered company address is: Gewerbestrasse 11, 6330 Cham, Switzerland

Supervisors' Foreword

Constructing a consistent fundamental quantum theory of gravity is often considered one of the most challenging open problems in modern physics. While a number of approaches are actively explored at present, none can yet claim complete success.

In particular, the idea of "Asymptotic Safety", originally due to Nobel Laureate Steven Weinberg, has attracted considerable attention over the past two decades. After the advent of modern functional techniques in the 1990s, there was mounting evidence hinting at the existence of a non-perturbatively renormalizable quantum field theory of the gravitational field. The corresponding framework makes essential use of Wilson's generalized notion of renormalization and renormalizability, proposing a strategy for circumventing the notorious non-renormalizability of quantized General Relativity in perturbation theory. According to Weinberg's idea, the limit of an infinite ultraviolet cutoff should be taken at a nontrivial renormalization group fixed point. This guarantees the absence of fatal divergences and is likely to lead to a quantum field theory which retains its predictive power at all scales. As a consequence, a main task of the Asymptotic Safety program consists in establishing the existence of a suitable fixed point and analyzing the properties of the renormalization group trajectories emanating from it. The latter contain the essential information about the dynamics predicted by the theory.

This thesis is devoted to an in-depth analysis of two aspects within the Asymptotic Safety program which are of central importance for the research field and beyond. The first aspect concerns the signature of the spacetime metrics that underlies the functional integral and the related RG equations. Almost all earlier investigations were based on fluctuations of the spacetime metric and assumed a Euclidean signature for technical simplicity. The present thesis develops an alternative formulation based on the Arnowitt–Deser–Misner (ADM)-decomposition of the metric which lends itself to a representation of both Lorentzian and Euclidean metrics, and under certain conditions even allows switching from one to the other (in the presence of a time-like Killing vector field). The construction of the renormalization group equation for the ADM-framework, successfully tested in this thesis, contains a considerable number of very creative and original conceptual developments and spearheads the development of a quantum theory of gravity incorporating Lorentzian spacetime signature. Particularly ingenious is the novel gauge-fixing scheme which it uses; it ensures a relativistic dispersion relation for all field components so that they propagate with the same speed. This property is likely to reduce the necessary quantum corrections to their physically required minimum, making the calculations particularly reliable and precise.

The second main topic of this thesis concerns the astrophysical and cosmological implications of Asymptotic Safety. In both cases, the method of "renormalization group improvement" is used in order to determine the leading quantum gravity corrections, exploiting the running couplings (Newton and cosmological constant) as the key input. The detailed discussion clarifies all steps leading to a viable low-energy effective action for inflation within the slow-roll framework. Besides the R^2-term characteristic for the Starobinsky Lagrangian, the resulting actions also contain additional curvature corrections expected from Asymptotic Safety. Fundamental questions regarding the fate of a star that undergoes a gravitational collapse are also discussed in connection with the Cosmic Censorship conjecture, so that the reader is introduced to many interesting facets of the phenomenological implications related to Asymptotic Safety.

The thesis is at the very forefront of modern quantum gravity research. Its exceptionally broad range of topics makes it appealing to a vast part of the international research community. We congratulate Dr. Platania on her technically outstanding and at the same time accessible work.

Catania, Italy   Dr. Alfio Bonanno
Nijmegen, The Netherlands   Dr. Frank Saueressig
June 2018

Abstract

General Relativity provides a remarkably successful description of gravity in terms of the geometric properties of spacetime. Nevertheless, the quantum nature of matter and the existence of regimes in which the classical description of gravity breaks down suggest that the short-distance behavior of spacetime may require a more fundamental quantum theory for the gravitational interaction. However, as it is well known, the quantization of General Relativity leads to a (perturbatively) non-renormalizable theory.

The Asymptotic Safety scenario for Quantum Gravity provides a natural mechanism for constructing a fundamental quantum theory for the gravitational force within the framework of Quantum Field Theory. In this scenario, the short-distance behavior of gravity is governed by a Non-Gaussian Fixed Point (NGFP) of the underlying renormalization group flow. The high-energy modifications of gravity resulting from the scaling of couplings around the NGFP may have profound implications in astrophysics and cosmology. The Functional Renormalization Group (FRG) constitutes an ideal tool to explore both the fundamental aspects and phenomenological implications of this scenario.

The aim of this Ph.D. thesis is twofold. First, we discuss the formulation of a Functional Renormalization Group Equation (FRGE) tailored to the Arnowitt–Deser–Misner (ADM) formalism. The construction also includes an arbitrary number of matter fields minimally coupled to gravity. This allows us to analyze the effect of matter on the fixed point structure of the gravitational renormalization group flow. Within a certain class of approximations, it will be shown that most of the commonly studied matter models, including the Standard Model of particle physics, give rise to an NGFP with real critical exponents. This result is important for the second part of this thesis, the phenomenological implications of Asymptotic Safety. Specifically, using a renormalization group improvement procedure, we study the quantum gravitational corrections arising in two different situations: the inflationary phase in the very early universe and the formation of black holes in the gravitational collapse of massive stars. In the context of cosmology, it will be shown that the predictions of Asymptotic Safety lead to an inflationary model compatible with the recent Planck data. In addition, a comparison between the inflationary models derived from foliated quantum gravity-matter systems with observations can put constraints on the primordial matter content of the universe. Finally, the study of gravitational collapse reveals that the anti-screening behavior of Newton's coupling in the short-distance limit renders the strength of the gravitational tidal forces weaker: the strong singularity appearing in the classical treatment is turned into a weak singularity once corrections from the renormalization group are included.

Acknowledgements

First and foremost, I would like to express my deepest and sincere gratitude to my scientific supervisors Alfio Bonanno (University of Catania) and Frank Saueressig (Radboud University Nijmegen) for their extraordinary scientific guidance, patience, and encouragement during my research and study.

I especially thank my main supervisor Alfio Bonanno for his inspirational guidance during all my doctoral studies. He has taught me how to do research and given me plenty of opportunities to improve my research skills, enter into new collaborations, and present my work at international conferences. I also thank my supervisor Frank Saueressig for his continuous support, his meticulous suggestions, and for showing perpetual confidence in my skills. He has always been willing to help me and answer all my questions. His energy and enthusiasm in research has given me the motivation to work harder and make the best effort possible.

Moreover, I wish to express my sincerest gratitude to Giuseppe Angilella for his constant guidance, availability, and valuable help during all my studies. In the course of my bachelor, master, and doctoral degrees I have learned and benefited enormously from his advises, his exceptional teaching capabilities, and remarkable support.

I would like to thank the University of Catania and Radboud University Nijmegen for giving me the opportunity to carry out my research project through a Double PhD program. In particular, I am very grateful to have been the recipient of the INFN Fellowship granted by the INFN (Catania Section) for the XXX–cycle Ph.D. Program of the University of Catania. I also thank INFN, INAF, and Radboud University Nijmegen for the financial support they provided me during my doctoral studies. This allowed me to attend various international conferences, workshops, and schools.

I am very thankful to Giuseppe Angilella, Alfio Bonanno, Fedele Lizzi, and Frank Saueressig for nominating and supplying my thesis to the *Springer Theses* program. In particular, I would like to thank Fedele Lizzi, Chair of my Doctoral Degree Committee, who acted as the official endorser of my dissertation. Finally,

I am very grateful to Springer, for recognizing my research and publishing my work in such a prestigious book series.

Last but not least, I wish to give my special thanks to my family for their unending love which has been with me all the time.

Contents

Part IV   Conclusions

Abbreviations

ADM	Arnowitt–Deser–Misner
AS	Asymptotic Safety
BFM	Background Field Method
CDT	Causal Dynamical Triangulation
CMB	Cosmic Microwave Background
EAA	Effective Average Action
ERGE	Exact Renormalization Group Equation
FRG	Functional Renormalization Group
FRGE	Functional Renormalization Group Equation
FRW	Friedmann–Robertson–Walker
GFP	Gaussian Fixed Point
GUT	Grand Unified Theory
IR	Infrared
MSSM	Minimal Supersymmetric Standard Model
NGFP	Non-Gaussian Fixed Point
QEG	Quantum Einstein Gravity
QG	Quantum Gravity
RG	Renormalization Group
SM	Standard Model
TT	Transverse-Traceless
UV	Ultraviolet
VKP	Vaidya–Kuroda–Papapetrou

Part I
Asymptotically
Safe Quantum Gravity

Chapter 1
Introduction

Quantum Field Theory is the standard framework for the description of the weak, electromagnetic and strong interactions, and results in an extremely well tested theory known as Standard Model (SM) of particle physics. Similarly, General Relativity provides a successful description of the gravitational interaction and most of its predictions have been confirmed by observations. Although Standard Model and General Relativity show a very good agreement with experimental observations, there are several inconsistencies and unsolved problems suggesting that these theories are incomplete and may not be able to describe all fundamental aspects of our universe. For instance, the SM cannot explain the observed baryon asymmetry characterizing the observable universe and cannot incorporate the neutrino masses, required to explain neutrino flavor oscillation, in a natural way. Moreover, most of the energy density filling our universe cannot be explained in terms of ordinary matter. In particular, the existence of "dark energy" governing the current phase of accelerated expansion constitutes a real conundrum. In principle it could be explained by a negative pressure associated with the vacuum energy of our universe (cosmological constant). However, the observed value is much smaller than the vacuum energy predicted by the Standard Model of particle physics, thus making the physical origin of dark energy completely obscure. In addition to the above issues, one of the most fundamental problems of General Relativity is the existence of regimes where the gravitational tidal forces blow up and result in the formation of spacetime singularities. Examples are the big bang in the very early universe and singularities associated with black holes. In these cases the impossibility of uniquely determining the evolution of the spacetime beyond the singularity marks the breakdown of the classical description of gravity within the context of General Relativity. On the other hand, the occurrence of such singularities is encountered in regimes where neither gravity nor quantum effects can be neglected. It is commonly believed that a quantum theory for the gravitational interaction may shed some light on these problems. A full fledged quantum theory of gravity in which these questions can be addressed is not yet available though.

© Springer Nature Switzerland AG 2018
A. B. Platania, *Asymptotically Safe Gravity*, Springer Theses,
https://doi.org/10.1007/978-3-319-98794-1_1

The formulation of a quantum theory for the gravitational interaction presents several conceptual problems and difficulties. In the framework of Quantum Field Theory the goal would be to combine the SM (on curved spacetimes) with a quantum causal theory of the spacetime, capable of describing the fundamental, microscopic aspects of the gravitational interaction. In this setting, quantum fluctuations of spacetime trigger an energy-scale-dependence of the gravitational couplings, just as quantum fluctuations of matter give rise to the running matter couplings of the SM. However, the quantization of General Relativity results in a (perturbatively) non-renormalizable theory, as the Newton's coupling has negative mass dimension and grows boundlessly towards high energies. A compelling solution to this problem, first proposed by Weinberg [1], is a generalization of the notion of renormalizability [2] based on the Wilsonian idea of renormalization [3]. The Wilsonian Renormalization Group (RG) relies on non-perturbative functional methods and naturally produces a scale-dependent effective action which interpolates between the fundamental microscopic theory and the low-energy effective dynamics. The set of trajectories "drawn" by the running couplings in the theory space is called RG flow. A quantum theory is then renormalizable and possesses a well-defined high-energy completion if the RG flow attains a fixed point in the ultraviolet (UV) limit [2]. In fact, an UV-attractive fixed point might govern the asymptotic high-energy behavior of the theory, thus preventing the appearance of unphysical divergences. In the case of a UV-attractive Gaussian Fixed Point (GFP), i.e. an asymptotically free theory, this generalized definition of renormalizability matches the perturbative one. The statement that General Relativity is (perturbatively) non-renormalizable means that the RG flow is repelled by the GFP in the high-energy regime. On the other hand, the gravitational RG flow might converge to a Non-Gaussian Fixed Point (NGFP) in the ultraviolet limit, thus rendering the theory "asymptotically safe". Such a non-trivial fixed point corresponds to an interacting theory and, as proposed by Weinberg [1], its existence would guarantee the renormalizability of gravity within the framework of Quantum Field Theory. The Asymptotic Safety conjecture can thus be summarized as follows: gravity is a finite and predictive Quantum Field Theory whose continuum limit is governed by a finite-dimensional critical surface in the theory space of diffeomorphism invariant metric theories. Accordingly, the quantization of gravity would result in a (non-pertubatively) renormalizable quantum theory whose high-energy completion is defined by a NGFP. This is called the Asymptotic Safety scenario for Quantum Gravity and the corresponding quantum theory is usually referred to as Quantum Einstein Gravity (QEG).

In recent years Asymptotically Safe Gravity has received much attention as several studies, employing the (non-perturbative) approach of the Functional Renormalization Group (FRG) [4–7], have established the presence of a NGFP suitable for Asymptotic Safety in a vast number of approximations [8–36]. An interesting consequence of Asymptotic Safety is that the original four-dimensional spacetime undergoes a dimensional reduction in the short-distance regime. Starting from the classical four-dimensional spacetime, realized at macroscopic length scales, the spectral dimension of the "emergent" effective spacetime varies with the energy scale and reaches the value $d_{\text{eff}} = 2$ in the ultraviolet limit [12]. Notably, the same result has been obtained in other approaches to Quantum Gravity, such as Hořava-Lifshitz gravity [37], Causal Dynamical Triangulation (CDT) [38], Loop Quantum Gravity (LQG) [39] and, quite recently, double special relativity [40]. This coincidence opens the possibility that all these approaches may describe different facets of the same quantum theory.

In the context of Asymptotic Safety, the quantization of gravity is achieved in the path integral formalism, where the action is specified in its Euclidean version and the gravitational degrees of freedom are encoded in the fluctuations of the (Euclidean) spacetime metric. This construction defines the so-called metric approach to Asymptotic Safety. The causal structure of the spacetime dictated by General Relativity is however encoded in a Lorentzian metric and it should be taken into account in order to consistently describe the fundamental aspects of quantum-relativistic theories. In the case of non-gravitational quantum theories, this issue can be circumvented because Lorentzian and Euclidean theories are related by a Wick rotation of the time coordinate. In the case of the gravitational interaction this problem turns out to be much more tricky and conceptual. Firstly, the interpretation of time in quantum mechanics and in General Relativity is different. In the former the flow of time is absolute and the time evolution of quantum states is determined by the Hamiltonian of the system. In the latter, time is relative and "hidden" in the spacetime metric. Moreover, the Hamiltonian of General Relativity must vanish to allow general covariance. This Hamiltonian constraint, in combination with the basic concepts of quantum mechanics, imply that time does not exist in General Relativity and results in the so-called "problem of time" [41]. Secondly, the spacetime metric is the dynamical field describing the gravitational interaction and, due to the general covariance of the theory, performing a Wick rotation is non-trivial. A quantum theory of gravity based on Euclidean computations may not be sufficient for the description of our universe, as the causal structure of the spacetime is not taken into account and cannot be recovered through the standard Wick rotation. A natural way to address the latter issue is the Arnowitt–Deser–Misner (ADM) formalism, in which the spacetime metric is decomposed into a stack of space-like surfaces, each one labeled by a given instant of time. The resulting distinguished time direction allows in principle the continuation of the flow equation from Euclidean to Lorentzian signature. In particular, recent studies based on the Matsubara-formalism have shown that the NGFP underlying Asymptotically Safe Gravity is stable under Wick rotation from Euclidean to Lorentzian signature [42, 43]. Moreover, the ADM-formalism in Quantum Gravity constitutes the natural bridge connecting the FRG to the CDT program [44].

The FRG approach to quantum gravity presents two important advantages: firstly, it permits to systematically analyze the non-perturbative renormalizability of gravity and to study formal aspects of the theory by means of non-perturbative methods. Secondly, it acts as a "microscope" capable of showing the properties of gravity at different length scales. In fact, the modern FRG techniques are based on the concept of the "effective average action", a scale-dependent effective action which smoothly interpolates between the short and long-distance regimes. The FRG thus provides an ideal tool to study astrophysical and cosmological implications of Asymptotic Safety. The quantum gravitational effects produced in proximity to the NGFP can in fact be encoded in a set of scale-dependent couplings. In particular, the RG-induced evolution of the Newton's coupling and cosmological constant could entail a natural and consistent "cosmic history" of our universe, from the initial Big Bang singularity to the current phase of accelerated expansion (see [45] for a comprehensive review).

This Ph.D. thesis focuses on two different fundamental aspects of Asymptotic Safety. Firstly, we will discuss the construction of Asymptotically Safe Gravity within the ADM-formalism, and we will initiate the study of gravity-matter systems on foliated spacetimes. Secondly, we will focus on some phenomenological implications of Asymptotic Safety.

In the present dissertation the formulation of a Functional Renormalization Group Equation (FRGE) within the ADM-formalism will be discussed. At variance of other related studies in the literature, our construction of the FRGE for foliated spacetimes does not require the time direction to be compact. As an application, we will study the renormalization group flow resulting from the ADM-decomposed Einstein–Hilbert truncation, evaluated on a $D = (d + 1)$-dimensional Friedmann–Robertson–Walker (FRW) background [46]. This construction will also include an arbitrary number of scalar, vector and Dirac fields minimally coupled to gravity [47]. The derivation of the beta functions for the cosmological constant and Newton's coupling will be reported and analyzed in detail. The resulting gravitational renormalization group flow will be studied and the fixed point structure induced by the presence of matter fields will be discussed. In addition, we will analyze the existence of fixed points for the pure gravitational flow as a function of the spacetime dimension [46]. Remarkably, the formalism we constructed constitutes a first important step towards connecting the FRG framework to the CDT program [48]. Furthermore, the study of foliated gravity-matter systems we propose in this dissertation employs the same fluctuation fields used in cosmology. This construction is thus perfectly suited to investigate the early universe cosmology within the context of Asymptotic Safety.

The recent observations of the cosmic microwave background (CMB) radiation [49] and the discovery of gravitational waves [50] have marked the beginning of a "golden era" for observational astrophysics and cosmology. In particular, observations on gamma ray bursts constitute an extremely important resource for investigating ultra-high energy phenomena. In these high-energy regimes, the quantum gravitational effects dynamically generated around the NGFP might drastically change the spacetime dynamics and could indirectly affect macroscopic details of various gravitational phenomena. Accordingly, the short-distance modifications of gravity induced by Asymptotic Safety could result in observable consequences. The Renormalization Group allows to describe the behavior of the gravitational interaction from the high-energy regime down to the low-energy world. The FRG thus provides a natural framework to investigate the astrophysical and cosmological implications of Asymptotic Safety and make predictions that can be compared with observations. Starting from a classical model, leading-order quantum gravitational effects can be taken into account by employing a renormalization group improvement procedure. The latter is a standard strategy developed in the context of QFT and it has been used successfully to study radiative corrections in scalar electrodynamics [51] and vacuum polarizations effects in both Quantum Electrodynamics and Quantum

Chromodynamics [52–56]. The RG improvement procedure thus constitutes a powerful tool to study the impact of quantum gravity effects on classical gravitational phenomena. In particular, the present dissertation focuses on the implications of Asymptotically Safe Gravity in two different scenarios: inflationary cosmology and black holes physics.

The short-distance modifications of gravity are expected to have an important impact in the primordial evolution of the universe. According to standard cosmology, the primordial quantum fluctuations were exponentially stretched during inflation, resulting in small density perturbations at the decoupling era. In the context of Asymptotically Safe Gravity these quantum fluctuations can be identified with the fluctuations of the spacetime geometry described by the NGFP regime [57]. The anisotropies in the CMB, due to the density fluctuations at the last scattering surface, can thus be traced back to the quantum gravity effects occurring during the Planck era. As the NGFP state corresponds to a scale-invariant theory, the scaling properties of the 2-point correlation function of the graviton around the NGFP induce a nearly scale-invariant spectrum of the primordial perturbations. Using an RG improvement procedure, a class of inflationary models arising from the short-distance modifications of pure gravity will be derived. A systematic comparison with the Planck data will be also be reported. Remarkably, our study shows that Asymptotic Safety predicts values for the spectral index in good agreement with the Planck data, and a tensor-to-scalar ratio which is significantly higher than the typical values obtained within classical inflationary models [58, 59]. This result is of crucial importance to test Asymptotic Safety. In fact, although the Planck data [49] provides only an upper limit for the tensor-to-scalar ratio, future experiments on the CMB anisotropies will provide more precise measurements [60–62]. We will then study a class of effective cosmological models emerging from quantum gravity-matter systems. By requiring the compatibility of these models with observational data, our analysis will provide important constraints on the matter content of the primordial universe [63].

As a final application of the Asymptotic Safety scenario for Quantum Gravity, we will discuss the problem of singularities in General Relativity and, more precisely, their formation during the gravitational collapse of a massive spherical star. Although the inclusion of the leading quantum corrections is not capable of removing the central singularity, it will be shown that the anti-screening character of the running Newton's coupling renders the classical singularity much milder ("gravitationally weak", according to the Tipler classification [64]), thus allowing the spacetime to be continuously extended beyond the singularity [65–67].

The present dissertation is organized in three parts. Part I summarizes the basics of functional renormalization. In particular, starting from the concept of universality in statistical physics, Chap. 2 provides a brief summary of the Wilsonian idea of renormalization. In particular, it introduces the generalized notion of renormalizability which is the starting point for investigating the Asymptotic Safety conjecture in Quantum Gravity. The latter is discussed in detail in Chap. 3, where the renormalization group equations are derived and some of the basic results of Functional Renormalization in Quantum Gravity are reviewed. The special case of the Einstein–Hilbert truncation is also discussed. Part II, and specifically Chap. 4, initiates our investigation of Quantum Gravity on foliated spacetimes. It provides a detailed exposition of the ADM-formalism in Asymptotic Safety and the derivation of the beta functions for foliated gravity-matter systems within the Einstein–Hilbert truncation [46–48]. Some technical details of the construction have been moved to Appendix A. The last part of this thesis, Part III, discusses astrophysical and cosmological implications of Asymptotic Safety. Chapter 5 is devoted to the study of the inflationary scenario arising from Asymptotically Safe Gravity [58, 59, 63]. The comparison of our theoretical predictions with the recent Planck data will also be reported. In Chap. 6 we will discuss the problem of the gravitational collapse and the formation of spacetime singularities in the framework of Asymptotically Safe theories of gravity [65–67]. At last, in Chap. 7, we will provide a summary and discussion of our findings.

References

1. S. Weinberg, Critical phenomena for field theorists, in *Proceedings 14th International School of Subnuclear Physics*, Erice (1976), p. 1. https://doi.org/10.1007/978-1-4684-0931-4_1 (cit. on p. 4)
2. K.G. Wilson, J. Kogut, The renormalization group and the ϵ expansion. Phys. Rep. **12**, 75–199 (1974). https://doi.org/10.1016/0370-1573(74)90023-4 (cit. on p. 4)
3. K.G. Wilson, Renormalization group and critical phenomena. I. Renormalization group and the Kadanoff scaling picture. Phys. Rev. B **4**, 3174–3183 (1971). https://doi.org/10.1103/PhysRevB.4.3174 (cit. on p. 4)
4. F.J. Wegner, A. Houghton, Renormalization group equation for critical phenomena. Phys. Rev. A **8**, 401–412 (1973). https://doi.org/10.1103/PhysRevA.8.401 (cit. on p. 5)
5. C. Wetterich, Exact evolution equation for the effective potential. Phys. Lett. B **301**, 90–94 (1993). https://doi.org/10.1016/0370-2693(93)90726-X (cit. on p. 5)
6. T.R. Morris, The exact renormalization group and approximate solutions. Int. J. Mod. Phys. A **9**, 2411–2449 (1994). https://doi.org/10.1142/S0217751X94000972. eprint: hep-ph/9308265 (cit. on p. 5)
7. M. Reuter, C. Wetterich, Effective average action for gauge theories and exact evolution equations. Nucl. Phys. B **417**, 181–214 (1994). https://doi.org/10.1016/0550-3213(94)90543-6 (cit. on p. 5)
8. M. Reuter, Nonperturbative evolution equation for quantum gravity. Phys. Rev. D **57**, 971–985 (1998). https://doi.org/10.1103/PhysRevD.57.971. eprint: hep-th/9605030 (cit. on p. 5)
9. W. Souma, Non-trivial ultraviolet fixed point in quantum gravity. Prog. Theor. Phys. **102**, 181–195 (1999). https://doi.org/10.1143/PTP.102.181. eprint: hep-th/9907027 (cit. on p. 5)
10. O. Lauscher, M. Reuter, Ultraviolet fixed point and generalized flow equation of quantum gravity. Phys. Rev. D **65**(2), 025013 (2002). https://doi.org/10.1103/PhysRevD.65.025013. eprint: hep-th/0108040 (cit. on p. 5)
11. M. Reuter, F. Saueressig, Renormalization group flow of quantum gravity in the Einstein–Hilbert truncation. Phys. Rev. D **65**(6), 065016 (2002). https://doi.org/10.1103/PhysRevD.65.065016. eprint: hep-th/0110054 (cit. on p. 5)
12. D.F. Litim, Fixed points of quantum gravity. Phys. Rev. Lett. **92**(20), 201301 (2004). https://doi.org/10.1103/PhysRevLett.92.201301. eprint: hep-th/0312114 (cit. on p. 5)
13. O. Lauscher, M. Reuter, Flow equation of Quantum Einstein Gravity in a higher-derivative truncation. Phys. Rev. D **66**(2), 025026 (2002). https://doi.org/10.1103/PhysRevD.66.025026. eprint: hep-th/0205062 (cit. on p. 5)

14. A. Codello, R. Percacci, C. Rahmede, Ultraviolet properties of f(R)-gravity. Int. J. Mod. Phys. A **23**, 143–150. https://doi.org/10.1142/S0217751X08038135. arXiv:0705.1769 [hep-th] (cit. on p. 5)
15. P.F. Machado, F. Saueressig, On the renormalization group flow of f(R)-gravity. Phys. Rev.D **77**, 124045 (2008). https://doi.org/10.1103/PhysRevD.77.124045. eprint:arXiv:0712.0445 (cit. on p. 5)
16. A. Codello, R. Percacci, C. Rahmede, Investigating the ultraviolet properties of gravity with a Wilsonian renormalization group equation. Ann. Phys. **324**, 414–469 (2009). https://doi.org/10.1016/j.aop.2008.08.008. arXiv:0805.2909 [hep-th] (cit. on p. 5)
17. K. Falls et al., Further evidence for asymptotic safety of quantum gravity. Phys. Rev. D **93**(10), 104022 (2016). https://doi.org/10.1103/PhysRevD.93.104022 (cit. on p. 5)
18. M. Demmel, F. Saueressig, O. Zanusso, RG flows of Quantum Einstein Gravity in the linear-geometric approximation. Ann. Phys. **359**, 141–165 (2015). https://doi.org/10.1016/j.aop.2015.04.018. arXiv:1412.7207 [hep-th] (cit. on p. 5)
19. A. Codello, R. Percacci, Fixed points of higher-derivative gravity. Phys. Rev. Lett. **97**(22), 221301 (2006). https://doi.org/10.1103/PhysRevLett.97.221301. eprint: hep-th/0607128 (cit. on p. 5)
20. D. Benedetti, P.F. Machado, F. Saueressig, Asymptotic safety in higher-derivative gravity. Mod. Phys. Lett. A **24**, 2233–2241 (2009). https://doi.org/10.1142/S0217732309031521. arXiv:0901.2984 [hep-th] (cit. on p. 5)
21. D. Benedetti, P.F. Machado, F. Saueressig, Taming perturbative divergences in asymptotically safe gravity. Nucl. Phys. B **824**, 168–191 (2010). https://doi.org/10.1016/j.nuclphysb.2009.08.023. arXiv:0902.4630 [hep-th] (cit. on p. 5)
22. F. Saueressig et al., Higher derivative gravity from the universal renormalization group machine, in *PoS EPS-HEP2011* (2011), p. 124. arXiv:1111.1743 [hep-th] (cit. on p. 5)
23. D. Benedetti, F. Caravelli, The local potential approximation in quantum gravity. J. High Energy Phys. **6**, 17 (2012). https://doi.org/10.1007/JHEP06(2012)017. arXiv:1204.3541 [hep-th] (cit. on p. 5)
24. M. Demmel, F. Saueressig, O. Zanusso, Fixed-functionals of three dimensional Quantum Einstein Gravity. J. High Energy Phys. **11**, 131 (2012). https://doi.org/10.1007/JHEP11(2012)131. arXiv: 1208.2038 [hep-th] (cit. on p. 5)
25. J.A. Dietz, T.R. Morris, Asymptotic safety in the f(R) approximation. J. High Energy Phys. **1**, 108 (2013). https://doi.org/10.1007/JHEP01(2013)108. arXiv:1211.0955 [hep-th] (cit. on p. 5)
26. M. Demmel, F. Saueressig, O. Zanusso, Fixed functionals in asymptotically safe gravity, in *Proceedings of the 13th Marcel Grossmann Meeting*, Stockholm, Sweden (2015), pp. 2227–2229. https://doi.org/10.1142/9789814623995_0404. arXiv:1302.1312 [hep-th] (cit. on p. 5)
27. J.A. Dietz, T.R. Morris, Redundant operators in the exact renormalisation group and in the f(R) approximation to asymptotic safety. J. High Energy Phys. **7**, 64 (2013). https://doi.org/10.1007/JHEP07(2013)064 (cit. on p. 5)
28. D. Benedetti, F. Guarnieri, Brans-Dicke theory in the local potential approximation. New J. Phys. **16**(5), 053051 (2014). https://doi.org/10.1088/1367-2630/16/5/053051. arXiv:1311.1081 [hep-th] (cit. on p. 5)
29. M. Demmel, F. Saueressig, O. Zanusso, RG flows of Quantum Einstein Gravity on maximally symmetric spaces. J. High Energy Phys. **6**, 26 (2014). https://doi.org/10.1007/JHEP06(2014)026. arXiv:1401.5495 [hep-th] (cit. on p. 5)
30. R. Percacci, G.P. Vacca, Search of scaling solutions in scalar-tensor gravity. Eur. Phys. J. C **75**, 188 (2015). https://doi.org/10.1140/epjc/s10052-015-3410-0. arXiv:1501.00888 [hep-th] (cit. on p. 5)
31. J. Borchardt, B. Knorr, Global solutions of functional fixed point equations via pseudospectral methods. Phys. Rev. D **91**(10), 105011 (2015). https://doi.org/10.1103/PhysRevD.91.105011. arXiv:1502.07511 [hep-th] (cit. on p. 5)
32. M. Demmel, F. Saueressig, O. Zanusso, A proper fixed functional for four-dimensional Quantum Einstein Gravity. J. High Energy Phys. **8**, 113 (2015). https://doi.org/10.1007/JHEP08(2015)113. arXiv:1504.07656 [hep-th] (cit. on p. 5)

33. N. Ohta, R. Percacci, G.P. Vacca, Flow equation for f(R) gravity and some of its exact solutions. Phys. Rev. D **92**(6), 061501 (2015). https://doi.org/10.1103/PhysRevD.92.061501. arXiv:1507.00968 [hep-th] (cit. on p. 5)

34. N. Ohta, R. Percacci, G.P. Vacca, Renormalization group equation and scaling solutions for f(R) gravity in exponential parametrization. Eur. Phys. J. C **76**, 46 (2016). https://doi.org/10.1140/epjc/s10052-016-3895-1. arXiv:1511.09393 [hep-th] (cit. on p. 5)

35. P. Labus, T.R. Morris, Z.H. Slade, Background independence in a background dependent renormalization group. Phys. Rev. D **94**(2), 024007 (2016). https://doi.org/10.1103/PhysRevD.94.024007. arXiv:1603.04772 [hep-th] (cit. on p. 5)

36. J.A. Dietz, T.R. Morris, Z.H. Slade, Fixed point structure of the conformal factor field in quantum gravity. Phys. Rev. D **94**(12), 124014 (2016). https://doi.org/10.1103/PhysRevD.94.124014. arXiv:1605.07636 [hep-th] (cit. on p. 5)

37. P. Horava, Spectral dimension of the universe in quantum gravity at a Lifshitz point. Phys. Rev. Lett. **102**(16), 161301 (2009). https://doi.org/10.1103/PhysRevLett.102.161301. arXiv:0902.3657 [hep-th] (cit. on p. 5)

38. J. Ambjrn, J. Jurkiewicz, R. Loll, The spectral dimension of the universe is scale dependent. Phys. Rev. Lett. **95**(17), 171301 (2005). https://doi.org/10.1103/PhysRevLett.95.171301. eprint: hep-th/0505113 (cit. on p. 5)

39. L. Modesto, Fractal spacetime from the area spectrum. Class. Quantum Gravity **26**(24), 242002 (2009). https://doi.org/10.1088/0264-9381/26/24/242002. arXiv:0812.2214 [gr-qc] (cit. on p. 5)

40. G. Amelino-Camelia et al., Planck-scale dimensional reduction without a preferred frame. Phys. Lett. B **736**, 317–320 (2014). https://doi.org/10.1016/j.physletb.2014.07.030. arXiv:1311.3135 [gr-qc] (cit. on p. 5)

41. C.J. Isham, Canonical quantum gravity and the problem of time (1992), pp. 0157–288. arXiv:gr-qc/9210011 [gr-qc] (cit. on p. 5)

42. E. Manrique, S. Rechenberger, F. Saueressig, Asymptotically safe Lorentzian gravity. Phys. Rev. Lett. **106**(25), 251302 (2011). https://doi.org/10.1103/PhysRevLett.106.251302. arXiv:1102.5012 [hep-th] (cit. on p. 6)

43. S. Rechenberger, F. Saueressig, A functional renormalization group equation for foliated spacetimes. J. High Energy Phys. **3**, 10 (2013). https://doi.org/10.1007/JHEP03(2013)010. arXiv:1212.5114 [hep-th] (cit. on p. 6)

44. R. Loll, Discrete Lorentzian quantum gravity. Nucl. Phys. B Proc. Suppl. **94**, 96–107 (2001). https://doi.org/10.1016/S0920-5632(01)00957-4. eprint: hep-th/0011194 (cit. on p. 6)

45. A. Bonanno, F. Saueressig, Asymptotically safe cosmology—a status report. Comptes Rendus Phys. **18**, 254–264 (2017). https://doi.org/10.1016/j.crhy.2017.02.002. arXiv:1702.04137 [hep-th] (cit. on p. 6)

46. J. Biemans, A. Platania, F. Saueressig, Quantum gravity on foliated spacetimes: asymptotically safe and sound. Phys. Rev. D **95**(8), 086013 (2017). https://doi.org/10.1103/PhysRevD.95.086013. arXiv:1609.04813 [hep-th] (cit. on pp. 6, 7, 9)

47. J. Biemans, A. Platania, F. Saueressig, Renormalization group fixed points of foliated gravity-matter systems. J. High Energy Phys. **5**, 93 (2017). https://doi.org/10.1007/JHEP05(2017)093. arXiv:1702.06539 [hep-th] (cit. on pp. 6, 9)

48. A. Platania, F. Saueressig, Functional renormalization group flows on Friedman–Lemaitre–Robertson–Walker backgrounds. Found. Phys. (2018). https://doi.org/10.1007/s10701-018-0181-0 arXiv:1710.01972 [hep-th] (cit. on pp. 7, 9)

49. Planck Collaboration et al., Planck 2015 results. XIII. Cosmological parameters. A&A **594**, A13 (2016). https://doi.org/10.1051/0004-6361/201525830. arXiv:1502.01589 (cit. on pp. 7, 8)

50. B.P. Abbott et al., Observation of gravitational waves from a binary black hole merger. Phys. Rev. Lett. **116**(6), 061102 (2016). https://doi.org/10.1103/PhysRevLett.116.061102. arXiv:1602.03837 [gr-qc] (cit. on p. 7)

51. S. Coleman, E. Weinberg, Radiative corrections as the origin of spontaneous symmetry breaking. Phys. Rev. D **7**, 1888–1910 (1973). https://doi.org/10.1103/PhysRevD.7.1888 (cit. on p. 7)

52. A.B. Migdal, Vacuum polarization in strong non-homogeneous fields. Nucl. Phys. B **52**, 483–505 (1973). https://doi.org/10.1016/0550-3213(73)90575-0 (cit. on p. 7)

53. D.J. Gross, F. Wilczek, Asymptotically free gauge theories. I. Phys. Rev. D **8**, 3633–3652 (1973). https://doi.org/10.1103/PhysRevD.8.3633 (cit. on p. 7)

54. H. Pagels, E. Tomboulis, Vacuum of the quantum Yang-Mills theory and magnetostatics. Nucl. Phys. B **143**, 485–502 (1978). https://doi.org/10.1016/0550-3213(78)90065-2 (cit. on p. 7)

55. S.G. Matinyan, G.K. Savvidy, Vacuum polarization induced by the intense gauge field. Nucl. Phys. B **134**, 539–545 (1978). https://doi.org/10.1016/0550-3213(78)90463-7 (cit. on p. 7)

56. S.L. Adler, Short-distance perturbation theory for the leading logarithm models. Nucl. Phys. B **217**, 381–394 (1983). https://doi.org/10.1016/0550-3213(83)90153-0 (cit. on p. 7)

57. A. Bonanno, M. Reuter, Entropy signature of the running cosmological constant. JCAP **8**, 024 (2007). https://doi.org/10.1088/1475-7516/2007/08/024. arXiv:0706.0174 [hep-th] (cit. on p. 8)

58. A. Bonanno, A. Platania, Asymptotically safe inflation from quadratic gravity. Phys. Lett. B **750**, 638–642 (2015). https://doi.org/10.1016/j.physletb.2015.10.005. arXiv:1507.03375 [gr-qc] (cit. on pp. 8, 9)

59. A. Bonanno, A. Platania, Asymptotically safe R + R^2 gravity, in *PoS CORFU2015* (2016), p. 159 (cit. on pp. 8, 9)

60. A. Kogut et al., The primordial inflation explorer (PIXIE): a nulling polarimeter for cosmic microwave background observations. JCAP **7**, 025 (2011). https://doi.org/10.1088/1475-7516/2011/07/025. arXiv:1105.2044 (cit. on p. 8)

61. T. Matsumura et al., Mission design of LiteBIRD. J. Low Temp. Phys. **176**, 733–740 (2014). https://doi.org/10.1007/s10909-013-0996-1. arXiv:1311.2847 [astro-ph.IM] (cit. on p. 8)

62. CORE Collaboration, F. Finelli et al., Exploring cosmic origins with CORE: inflation. ArXiv e-prints (2016). arXiv:1612.08270 (cit. on p. 8)

63. A. Bonanno, A. Platania, F. Saueressig, Cosmological bounds on the field content of asymptotically safe gravity-matter models. Phys. Lett. B (2018). https://doi.org/10.1016/j.physletb.2018.06.047. arXiv:1803.02355 [gr-qc] (cit. on pp. 8, 9)

64. F.J. Tipler, On the nature of singularities in general relativity. Phys. Rev. D **15**, 942–945 (1977). https://doi.org/10.1103/PhysRevD.15.942 (cit. on p. 8)

65. A. Bonanno, B. Koch, A. Platania, Cosmic censorship in quantum Einstein gravity. Class. Quantum Gravity **34**(9), 095012 (2017). https://doi.org/10.1088/1361-6382/aa6788. arXiv:1610.05299 [gr-qc] (cit. on pp. 8, 9)

66. A. Bonanno, B. Koch, A. Platania, Asymptotically safe gravitational collapse: Kuroda-Papapetrou RG-improved model, in *PoS CORFU2016* (2017), p. 058 (cit. on pp. 8, 9)

67. A. Bonanno, B. Koch, A. Platania, Gravitational collapse in Quantum Einstein Gravity. Found. Phys. (2018). https://doi.org/10.1007/s10701-018-0195-7. arXiv:1710.10845 [gr-qc] (cit. on pp. 8, 9)

Chapter 2
The Wilsonian Idea of Renormalization

In the early decades of the 20th century, many investigations in the context of Quantum Field Theory showed that quantum-relativistic theories were affected by unphysical ultraviolet divergences. For instance, a systematic study of scattering amplitudes in Quantum Electrodynamics show that, due to the singular high-energy behavior of the theory, Feynman diagrams containing loops result in an infinite contribution to the transition amplitude.

The concept of renormalization was initially introduced in the context of Quantum Electrodynamics as a mathematical trick to solve the problem of infinities. The main idea was to absorb infinities into a redefinition of the parameters of the theory (bare couplings). For a renormalizable theory this procedure leads to a new effective theory where the bare couplings are replaced by a *finite* number of "renormalized" coupling constants [1], whose values can be fixed by experiments. As a consequence, the physical coupling constants depend on the energy scale at which they are measured. In particular, an equation describing the variation of the electric charge with the energy scale was devised by Gell-Mann and Low in 1954 [2] and can be considered as the first Renormalization Group equation.

In the same years, investigations of critical phenomena in condensed matter physics showed how different physical systems behave the same in the proximity of the critical point. A possible explanation for this phenomenon was suggested by Kadanoff in 1966 [3]: near the critical point the correlation length of the system blows up and the system "looks the same on all the length scales". This idea represented the starting point for the development of the Renormalization Group theory.

In the 1970s K. Wilson, inspired by the Gell-Mann and Low equation and the recent studies of Kadanoff, realized that in both quantum and statistical field theories fluctuations (quantum or thermal) play an important role, as they can explain both the renormalization effects and the universality properties of critical phenomena [4]. This idea led Wilson [5], as well as Kadanoff and Fisher [6, 7], to develop a set of non-perturbative techniques aimed at investigating the role of fluctuations in both quantum and thermal field theories, resulting in the modern formulation of the Renormalization Group.

© Springer Nature Switzerland AG 2018
A. B. Platania, *Asymptotically Safe Gravity*, Springer Theses,
https://doi.org/10.1007/978-3-319-98794-1_2

In this chapter we shall review the Wilsonian idea of renormalization and its implementation via non-perturbative functional techniques.

2.1 Scale Invariance, Universality and Kadanoff Blocking Transformations

Most of the laws of nature can be cast in the form of scaling laws. Such relations can generally be obtained by a dimensional analysis, and so the power index is expected to be a rational number. This number represents, in some sense, a "classical" scaling exponent. There is however a large class of phenomena, including *continuous phase transitions*, in which the characteristic exponents are real numbers and the scaling laws cannot be explained by a simple dimensional argument. In this section we will see how this feature is related to the lack of a characteristic length (or energy) scale, and how Kadanoff used this hypothesis to explain the universality properties of phase transitions.

2.1.1 The Role of Scaling Laws

In order to introduce the problem of critical phenomena and proceed with the study of the Wilsonian Renormalization Group, we first need to set up some notation and terminology. Let us consider a physical system (continuous or on a lattice) whose behavior depends upon N parameters $\{g_i\}$. The *phase space* is defined as the N-dimensional space whose points $(g_1, \ldots, g_N)$ represent all possible physical configurations of the system. Therefore, for a given initial condition, the dynamical evolution of the system can be visualized as a trajectory in the phase space. In the context of thermal field theory, a fundamental question is whether or not physical quantities are well defined along these trajectories. A *phase transition* is indeed associated with a singular behavior in the derivatives of the *bulk free energy density* $f_b(g_i)$ (free energy per unit of volume or per site) and the corresponding singular loci of the phase space can have any dimensionality $D_s \leq N$. These non-analyticity regions can be classified according to their co-dimension $C = N - D_s$, and the *phase boundaries*[1] are precisely the singular loci having co-dimension $C = 1$.

Phase transitions can be divided into two classes. First order phase transitions take place when a singular behavior of $f_b(g_i)$ occurs in its first derivative, while a second order or *continuous phase transition* is realized when the non-analyticity shows up

---

[1]Different phases are generally separated by the phase boundaries, but a change of phase does not necessarily involve a phase transition. In principle two different phases can be "connected" without passing through a phase boundary. In the latter case the two phases share the same degree of symmetry and it is not clear how these phases can actually be distinguished (see [8] for an extensive discussion).

only in higher derivatives. The dynamics of the system near the singular loci of the phase diagram constitutes a *critical phenomenon*.

Among the variety of critical phenomena, the liquid-gas and ferromagnetic phase transitions constitute the most important examples. Both these systems are characterized by a continuous phase transition at a finite (critical) value of the temperature, T_c. Continuous phase transitions can be modeled by using a "control parameter", for example the reduced temperature $t = |T - T_c|/T_c$, and an *order parameter* which allows to distinguish among different phases. In particular, a non-zero order parameter means that the system has reached a long-range ordered state. One of the most important feature of critical phenomena is that physical quantities exhibit "scaling" in proximity to the phase transition, namely they are related to each others through simple power laws. For instance, the physical law governing the dependence of the order parameter $\mathcal{O}$ upon the temperature can be cast in the form

$$\mathcal{O} \sim |T - T_c|^\beta, \qquad \beta \in \mathbb{R}. \tag{2.1}$$

The characteristic exponent β is an example of a *critical exponent*, and it turns out that almost all thermodynamical quantities follow a similar behavior in the vicinity of the critical point. Furthermore, even if ferromagnetic-paramagnetic and liquid-gas phase transitions are characterized by different order parameters, the corresponding characteristic exponents β are identical within the experimental error bars. In these cases, the value of this critical exponent was found to be $\beta \sim 0.325$ [9, 10]. This evidence was one of the first experimental proofs of the *universality properties* of systems at criticality, which is perhaps the most important treat of phase transitions. In particular, all systems having a Curie (critical) point show similar scaling properties in proximity to the critical region and systems characterized by the same set of critical exponents are said to be part of the same *universality class*.

Within the context of critical phenomena, the theoretical investigations of liquid-gas and ferromagnetic systems were initiated by Van der Waals and Weiss respectively. Their studies, based on a Mean Field Theory (MFT) approach, resulted in a good qualitative description of phase transitions, but they could neither predict the correct critical exponents nor explain the universal properties of such systems. The key problem was that their analyses were based on a "classical" description of the phenomena, in the sense that (thermal) fluctuations were not taken into account. The Mean Field Theory approximation is by construction an "effective" description of the theory, in which fundamental physical quantities are replaced with average fields. The reason why MFT failed to predict the correct critical exponents was indeed originated from its basic assumptions. As suggested by Wilson, a phase transition involves the interplay of fluctuations on all length scales, and hence, fluctuations constitute the key ingredient to understand the mechanism underlying critical phenomena. The failure of the Mean Field Theory approach and the subsequent development of Kadanoff's scaling hypothesis represented a fundamental step towards the development of the Wilsonian Renormalization Group.

2.1.2   Kadanoff Blocking Transformations

In order to introduce the Kadanoff scaling hypothesis, let us consider a spin system with lattice spacing a, described by the nearest neighborhood Ising model Hamiltonian

$$\frac{H[\sigma]}{k_B T} = -h \sum_{i=1}^{N} \sigma(\mathbf{r}_i) - g \sum_{\langle ij \rangle} \sigma(\mathbf{r}_i) \sigma(\mathbf{r}_j) \tag{2.2}$$

where $\mathbf{r}_i \equiv (ma, na, la)$, with $m, n, l \in \mathbb{Z}$, is a vector on the lattice which uniquely identifies the spins degrees of freedom $\sigma_i \equiv \sigma(\mathbf{r}_i)$, g is a dimensionless coupling constant and h is a dimensionless source (an external magnetic field, for instance). The idea underlying Kadanoff blocking transformations is to introduce a real number $l > 1$, and then study a new effective system with Nl^{-d} *spin blocks* variables, where d is the dimension of the system. These effective degrees of freedom are defined on a new hypercubic domain with lattice spacing $a_l = la$, and given by

$$\sigma_l(\mathbf{r}_I) = \frac{1}{l^d} \sum_{i \in I} \sigma(\mathbf{r}_i) \tag{2.3}$$

where the above sum extends over the original spin variables located within the block I. The original degrees of freedom are thus replaced by a set of effective variables describing the system with a smaller resolution. This procedure is known as *blocking transformation*.

The first fundamental assumption is that

$$a_l \ll \xi(T) \tag{2.4}$$

where the correlation length $\xi(T)$ is defined as the average extent of fluctuations about the mean value. In fact, by deriving the the two-point correlation function in the Ornstein-Zernike form

$$G(r = |\mathbf{r}_i - \mathbf{r}_j|) = \frac{e^{-r/\xi(T)}}{r^{d-2}} \xrightarrow[T \to T_c]{} \qquad G(r) \simeq r^{-(d-2+\eta)} \tag{2.5}$$

it is clear that ξ represents a measure of the correlation range of the system. This characteristic range depends upon the parameters of the theory, in particular the reduced temperature t. In proximity to the critical temperature the correlation length blows up and diverges as $t^{-\nu}$ once the critical point is reached.[2] In this limit the canonical scaling dimension of the correlation function gains an extra contribution η, which is usually referred to as *anomalous dimension*.

---

[2]It is important to remark that in principle continuous phase transitions can take place only in the thermodynamic limit. In fact, if the linear size of the system is finite, the correlation length cannot grow beyond the size of the system and so, in principle, the phase transition is smoothed out.

Kadanoff suggested that, approaching the critical point, the correlation length should become the only relevant length scale of the system. In particular, as $\xi(t) \to \infty$ at the critical temperature, the system is invariant under scale transformations and hence the Hamiltonian itself must remain unchanged (in form). Under this assumption, the blocking transformation does not modify the original symmetries, and thereby coupling constants, sources and the reduced temperature itself must be homogeneous functions under such transformations

$$t_l = t\, l^{y_t}, \qquad h_l = h\, l^{y_h}, \qquad g_l = g\, l^{y_g}. \tag{2.6}$$

Kadanoff's scaling hypothesis thus made it possible to explain the scaling behavior of physical quantities at criticality. Despite the fact that the blocking transformation does not allow to compute critical exponents nor explains universality, it gave important insights for the formulation of the Renormalization Group.

2.2 The Wilsonian Idea of Renormalization

In this section we will introduce the idea underlying the Renormalization Group in the intuitive framework of discrete dynamical systems.

The main question Wilson wanted to address in his work was what happens under repeated blocking transformations. Since fluctuations at criticality do not have a characteristic length scale, the very first insight of Wilson was to use dimensionless quantities from the beginning. Furthermore, as critical behavior involves infinitely many length scales, Wilson thought was to divide the whole problem into a "sequence of subproblems" [11]. Each subproblem does have a characteristic length scale and corresponds to a single step of the renormalization procedure. The starting point is a general Hamiltonian

$$\mathcal{H}[\sigma] = \frac{H[\sigma]}{k_B T} = -\sum_n g_n\, \Theta_n[\sigma] \tag{2.7}$$

where Θ_n are local operators in the spin variables $\sigma_i = \sigma(\mathbf{r}_i)$ and g_n are the corresponding *dimensionless* coupling constants. The main idea underlying the Renormalization Group (RG) mechanism is to combine a blocking transformation with a rescaling of the length unit by a factor l. This *dilatation* makes the system identical to the original one in terms of its domain and degrees of freedom, while the new Hamiltonian can be much more complicated. Such combined transformations take the name of Renormalization Group transformations and can be described by an abstract (non-linear) operator $\mathcal{R}_l$ which maps the original set of couplings $\mathbf{g} \equiv \{g_i\}$ into

$$\mathbf{g}_l = \mathcal{R}_l \cdot \mathbf{g} \tag{2.8}$$

where $\mathbf{g}_l$ are the new *renormalized* couplings describing the system at the renormalization scale l. In particular, the set of transformations $\mathcal{R}_l$ constitute a semi-group.

In fact, once the original degrees of freedom have been averaged out it is impossible to restore them back and so the transformations $\mathcal{R}_l$ cannot be inverted. In other words the length scale l must be $l \geq 1$, with $l = 1$ corresponding to the identity transformation. It is worth mentioning that there is no unique way to perform the coarse-graining, and different RG recursion relations may be employed depending on the problem at hand.

After a single RG transformation has been performed, the system is described by $N_l = N l^{-d}$ degrees of freedom whose dynamics is governed by an effective Hamiltonian $\mathcal{H}_l$. The latter naturally emerges by summing over the original dynamical variables and requiring that the partition function $\mathcal{Z}$ remains unchanged

$$e^{-\mathcal{H}_l[\sigma_l]} = \mathrm{Tr}_{\{\sigma\}} \left(\mathcal{P}[\sigma_l(\mathbf{r}_l), \sigma(\mathbf{r}_i)] e^{-\mathcal{H}[\sigma]} \right) \tag{2.9}$$

where the projection operator $\mathcal{P}[\sigma_l, \sigma]$ is a positive definite function such that

$$\mathrm{Tr}_{\{\sigma_l\}} \mathcal{P}[\sigma_l(\mathbf{r}_l), \sigma(\mathbf{r}_i)] = 1 \quad \Rightarrow \quad \mathcal{Z}_l \equiv \mathrm{Tr}_{\{\sigma_l\}} e^{-\mathcal{H}_l[\sigma_l]} \equiv \mathcal{Z} \tag{2.10}$$

and it depends upon the particular renormalization scheme (definition of the new blocks). As the projection operator acts non-linearly on $\mathcal{H}$, the Renormalization Group iteration can generate all local operators compatible with the original symmetries. The phase space to be considered is thus the *whole theory space* of Hamiltonians compatible with a given symmetry group.

A very important aspect of this step-by-step procedure relies in the analyticity of each RG transformation. The singular behavior arising in phase transitions can only emerge once all (wavelengths) fluctuations are taken into account, namely after the RG transformation has been repeated an infinite number of times. Hence, the critical behavior should result as a possible asymptotic ($l \rightarrow \infty$) state of the system. The goal of this procedure is then to find all possible asymptotic behaviors of Hamiltonians belonging to the same symmetry group.

At this stage it is important to make the connection between the Renormalization Group iteration and the phase diagram of the system. The original system can be represented as a point in the theory space whose coordinates are the couplings $\mathbf{g} \equiv \{g_i\}$. The RG transformation $\mathcal{R}_l$ can be iteratively applied and each step relates an initial point $\mathbf{g}_l^{(m)}$ to a new point $\mathbf{g}_l^{(m+1)}$ in the theory space. These points, or intermediate theories, are related to each others through recursion relations in the form

$$\mathbf{g}_l^{(n)} = \mathcal{R}_l \cdot \mathbf{g}_l^{(n-1)} = \mathcal{R}_l^2 \cdot \mathbf{g}_l^{(n-2)} = \cdots = \mathcal{R}_l^n \cdot \mathbf{g} \tag{2.11}$$

where, as a result of the semi-group properties of $\mathcal{R}_l$, $\mathbf{g}_l^{(n)} \equiv \mathbf{g}_{l^n}$. Therefore, the whole set of iterations traces out a trajectory in the phase space, whose initial condition is the original system described by the couplings $\mathbf{g}$. The family of all RG trajectories obtained by varying the initial conditions in the theory space constitutes the *renormalization group flow*.

It is natural to think that after a sufficiently large number of iterations the renormalization group flow will converge to an equilibrium stationary solution which represents the macroscopic state of the system. Such a solution is called a *fixed point* and is defined by the condition

$$\mathbf{g}_* = \mathcal{R}_l \cdot \mathbf{g}_*. \tag{2.12}$$

The set of couplings satisfying the above relation identify all possible asymptotic behaviors of the system, i.e. all possible endpoints of the RG trajectories. In principle the solutions of (2.12) can be loci of any dimensionality (fixed points, limit cycles, strange attractors), but generically one finds only fixed points. In particular, Eq. (2.12) can have several fixed point solutions, which play the role of infrared (long-distance) or ultraviolet (high-energy) attractors for the renormalization group flow. The whole theory space is thus divided into a set of *basins of attraction*, each one defined as the set of initial conditions whose renormalization group flow is attracted by a given fixed point in the long-distance limit.

A first classification of fixed points comes from the behavior of the correlation length. Since for each RG transformation the relevant length units are rescaled by the dilatation parameter l, the correlation length itself must follow the scaling law $\xi(\mathbf{g}_l^{(n)}) = \xi(\mathbf{g}_l^{(n-1)})/l$. It follows that the correlation length at the fixed point must satisfy

$$\xi(\mathbf{g}_*) = \xi(\mathbf{g}_*)/l \tag{2.13}$$

and hence it can only take the values zero or infinity. Infrared fixed points with $\xi(\mathbf{g}_*) = 0$ correspond to possible *bulk phases* of the system. Such solutions basically describe non-interacting theories and are usually referred to as "trivial" or Gaussian solutions. On the other hand, a fixed point solution for which $\xi(\mathbf{g}_*) \to \infty$ is called *critical fixed point*. It is the existence of such critical points that causes universality. The basin of attraction of a critical fixed point is also called *critical manifold* and represents the set of initial conditions which flow towards criticality in the infrared limit. Fixed points can be further classified according to the co-dimension of the corresponding basin of attraction. In particular, critical and multicritical manifolds have co-dimension $C \geq 2$ [8].

In order to describe the renormalization of couplings as a dynamical system, Eq. (2.8) must be put in differential form. Let us consider an RG transformation associated with a scale s

$$\mathbf{g}_{r=ls} = \mathcal{R}_s \cdot \mathbf{g}_l. \tag{2.14}$$

The question we need to answer is what happens under an infinitesimal transformation. Such a transformation is associated with an infinitesimal change of scale $s \equiv 1 + \varepsilon$ and differs from the identity just for a small quantity

$$\mathcal{R}_s \simeq \mathbb{1} + \left.\frac{\partial \mathcal{R}_s}{\partial s}\right|_{s=1} \varepsilon \tag{2.15}$$

where $\partial_s \mathcal{R}_s|_{s=1}$ is a matrix valued function acting on the vectors $\mathbf{g}_l$ of the theory space. The left-hand-side of Eq. (2.14) can thus be written as

$$\mathbf{g}_r \simeq \mathbf{g}_l + \frac{\partial \mathbf{g}_r}{\partial r}\bigg|_{r=l} \varepsilon l \tag{2.16}$$

and, on substituting the latter two equations into (2.14), one finds

$$l \partial_l \mathbf{g}_l = B(\mathbf{g}_l), \tag{2.17}$$

with $B(\mathbf{g}_l) = \partial_s \mathcal{R}_s|_{s=1} \cdot \mathbf{g}_l$. According to the above expression, the dynamics of the couplings in the theory space is described by a continuous trajectory $\mathbf{g}_l(\tau)$ parametrized by the "RG time"[3] $\tau = \log l$. The fixed points are thus the stationary solutions of Eq. (2.17), $B(\mathbf{g}_*) = 0$. The behavior of the flow in the proximity of a fixed point can be obtained by linearizing Eq. (2.17) about $\mathbf{g}_*$

$$\partial_\tau \delta \mathbf{g}(\tau) \simeq \mathcal{S}(\mathbf{g}_*) \delta \mathbf{g}(\tau) \tag{2.18}$$

where $\delta \mathbf{g}(\tau) \equiv \mathbf{g}(\tau) - \mathbf{g}_*$ and the *stability matrix* $\mathcal{S}_{ij} \equiv \partial_{g_j} B_i$ has been introduced. Denoting the eigenvalues of the stability matrix by y_n and the corresponding eigenperturbations by $\hat{y}_n$, the function $\delta \mathbf{g}(\tau)$ can be expanded as

$$\delta \mathbf{g}(\tau) = \sum_n \lambda_n(\tau) \hat{y}_n. \tag{2.19}$$

Thus the evolution of $\delta \mathbf{g}(\tau)$ is completely determined by the functions $\lambda_n(\tau)$. By substituting the above expansion into Eq. (2.18), one finds

$$\lambda_n(\tau) = \lambda_n(0) e^{y_n \tau}. \tag{2.20}$$

At this point it is of central importance to discuss the properties of eigenperturbations and eigenvalues. Eigenoperators can be firstly classified according to their physical relevance. Eigenvectors whose eigenvalues depend on the particular renormalization scheme do not affect physics and are called *redundant* [12]. On the contrary, if the characteristic exponents are invariant under changes of the renormalization scheme, the corresponding eigenperturbations are referred to as *scaling operators* [12]. On this basis, only the scaling operators are associated with the universality properties of physical systems, while redundant operator do not contribute to the critical behavior [12]. Scaling operators can be further classified as follows. Operators whose eigenvalues satisfy $\text{Re}(y_n) > 0$ are called relevant, and $\lambda_n(\tau)$ grows with τ. Vice versa, if $\text{Re}(y_n) < 0$ then $\lambda_n(\tau)$ decreases with τ and the corresponding operator is

---

[3]The appearance of the RG time $\tau = \log l$, which is a non-linear function of the length scale, is due to the multiplicative character of the RG transformations.

called irrelevant. The limiting case $\text{Re}(y_n) = 0$ corresponds to a marginal scaling operator if $\text{Im}(y_n) = 0$, and in this case $\lambda_n(\tau)$ is constant.

For a given fixed point, the number of relevant operators defines the co-dimension of its domain of attraction and so it establishes the nature of the fixed point in terms of criticality. As long as $\xi \to \infty$, the behavior of the RG trajectories near a critical fixed point is completely determined by the critical exponents and does not depend upon the microscopic details of the theory. Therefore, the universal scaling observed in critical phenomena is due to the dynamics of the renormalization group flow in the vicinity of a critical fixed point. A *universality class* can thus be interpreted as the basin of attraction of a critical fixed point: although the theories belonging to the same universality class may differ microscopically (different set of irrelevant parameters), they share the same set of relevant operators in the infrared (long-distance) limit and thus such theories behave the same at criticality.

2.3 Functional Renormalization Group

In the previous section we have seen how Renormalization Group transformations can be implemented within the context of lattice models. In the spirit of extending these ideas to systems with an infinite number of degrees of freedom, the path integral formalism is the key ingredient. In fact, as we shall see, the functional formulation of field theories provides the perfect framework to study renormalization non-perturbatively. The Wilsonian Renormalization Group complemented by these non-perturbative methods resulted in the so-called *Functional Renormalization Group* (FRG). The key idea is to introduce an infrared cutoff scale k and construct a generalized scale-depend action which smoothly interpolates between the bare action (high-energy regime) and the ordinary effective action (infrared limit). The FRG allows to deeply understand what renormalization really means and, as we shall see, it furnishes a natural generalization of the notion of renormalizability.

2.3.1 Wilsonian Average Action

Let us consider a field theory of a single scalar field whose *bare action* $S_\Lambda[\phi]$ is defined at the ultraviolet (UV) cutoff scale Λ. This UV cutoff plays the role of a microscopic minimal length $a \simeq \Lambda^{-1}$. The original scalar field $\phi(x)$ is thus defined on a spacetime lattice with lattice spacing Λ^{-1}. For the moment it is convenient to keep Λ finite, while the limit $\Lambda \to \infty$ can eventually be considered at the end of the calculation.

As the renormalization of a Quantum Field Theory is usually carried out in momentum space, we first introduce an arbitrary energy scale $k < \Lambda$ which plays the role of an *infrared cutoff*. The aim of this subsection is to find the general form of the Wilsonian action at the scale k. As we have seen, a blocking transformation consists of a redefinition of the bare degrees of freedom in terms of new effective variables describing the system at smaller resolutions. In the case of field theories one may follow the same strategy [11, 13]. The "block fields" $\phi_k(x)$ can be defined by averaging over spacetime blocks of linear dimensions $k^{-1} > \Lambda^{-1}$. The corresponding effective interactions can be obtained by integrating out the fluctuating modes with momenta $q \in [k, \Lambda[$. All information corresponding to these fast-fluctuations is thus "hidden" in averaged fields interacting through the effective action at the scale k.

Let us consider a finite region of the spacetime $\mathcal{B}_k(x)$ having linear dimension k^{-1} and volume $\Omega(\mathcal{B}_k)$. The average field $\phi_k(x)$ can be defined as follows

$$\phi_k(x) = \frac{1}{\Omega(\mathcal{B}_k)} \int_{\mathcal{B}_k(x)} \phi(y)\, d^D y = \int \rho_k(x - y)\phi(y)\, d^D y \qquad (2.21)$$

where the smearing function $\rho_k(x - y)$ vanishes when y falls outside the block, i.e. for $|x - y| \gtrsim k^{-1}$. A Renormalization Group transformation modifies the fundamental components and interactions of the system, but it must leave its long-distance properties, i.e. the averaged quantities, unchanged. Then one must require

$$\mathcal{Z} = \int \mathcal{D}[\phi]\, e^{-S_\Lambda[\phi]} \equiv \int \mathcal{D}[\Phi]\, e^{-S_k[\Phi]}. \qquad (2.22)$$

Here the partition function $\mathcal{Z}$ is written in terms of a Euclidean field integral and Φ denotes the new average field at the scale k. Given the similarities of quantum and thermal systems under RG transformations, it is natural to define the k-dependent effective action $S_k[\Phi]$ as

$$e^{-S_k[\Phi]} = \int \mathcal{D}[\phi]\, \mathcal{P}[\Phi(x), \phi(x)] e^{-S_\Lambda[\phi]} \qquad (2.23)$$

where the projection operator $\mathcal{P}[\Phi, \phi]$ plays the same role as the one introduced in Eq. (2.9) and shares similar properties. The action $S_k[\Phi]$ defined above is known as *Wilsonian action*. This generalized action describes the physics at the scale k and includes all (quantum or thermal) fluctuations at higher energy scales. In addition, the action $S_k[\Phi]$ is a continuous function of the infrared cutoff k. The bare and effective action are indeed recovered in the ultraviolet and infrared limit respectively.

With the purpose of finding an explicit expression for $S_k[\Phi]$, the Fourier modes $\phi(q)$ with momenta $q \gtrsim k$ must be integrated out. This integration can be performed by employing a sharp cutoff function

$$\rho_k(q) = \theta(k - q). \qquad (2.24)$$

This choice for the smearing function provides a natural splitting of $\phi(x)$ into the slow and fast varying contributions, $\phi(x) = \phi_k(x) + \xi(x)$. In the momentum space this splitting reads

$$\phi(q) = \begin{cases} \phi_k(q) & q \in [0, k[\\ \xi(q) & q \in [k, \Lambda[. \end{cases} \qquad (2.25)$$

Hence, the Fourier components $\phi_k(q)$ define the effective background field $\phi_k(x)$, whilst $\xi(x)$ corresponds to the fluctuations around this background. In order to obtain the action functional at the scale k, we need to remove these fluctuations by integrating out, shell by shell, all Fourier modes from the ultraviolet cutoff Λ to the (variable) infrared scale k. The Fourier components $\xi(q)$ are thus the fields over which the integration will be performed. In addition, we require that the average field $\Phi(x)$ in Eq. (2.23) coincides with the background field $\phi_k(x)$ defined above. Therefore the projection operator is constrained to be a delta functional

$$\mathcal{P}[\Phi(x), \phi(x)] \equiv \prod_x \delta[\phi_k(x) - \Phi(x)]. \qquad (2.26)$$

Following the arguments in [13–15], we now expand $S_\Lambda[\phi]$ around $\xi = 0$

$$S_\Lambda[\phi] = S_\Lambda[\phi_k] + \int_k^\Lambda \left\{ S_\Lambda^{(1)}[\phi_k]\xi(q) + \tfrac{1}{2}\xi(q) S_\Lambda^{(2)}[\phi_k]\xi(-q) \right\} \frac{d^D q}{(2\pi)^d} + \cdots \qquad (2.27)$$

and thus

$$e^{-S_k[\Phi]} \simeq \int [D\phi_k]\,\delta(\phi_k - \Phi) \int [D\xi]\, e^{-S_\Lambda[\phi_k] - \tfrac{1}{2}\xi(q)\cdot S_\Lambda^{(1)}[\phi_k] - \tfrac{1}{2}\xi(q)\cdot S_\Lambda^{(2)}[\phi_k]\cdot\xi(-q)+\cdots} \qquad (2.28)$$

where the de Witt notation has been used. Performing the Gaussian integral, the Wilsonian action $S_k[\Phi]$ finally reads

$$S_k[\Phi] \simeq S_\Lambda[\Phi] + \tfrac{1}{2}\,\mathrm{STr}'\left[\log\left(S_\Lambda^{(2)}[\Phi]\right)\right] - \tfrac{1}{2}\,\mathrm{Tr}'\left[S_\Lambda^{(1)}[\Phi]\left(S_\Lambda^{(2)}[\Phi]\right)^{-1} S_\Lambda^{(1)}[\Phi]\right]. \qquad (2.29)$$

Here the trace Tr' stands for an integration over the momenta $q \in [k, \Lambda[$. In addition, the super trace STr' also includes a sum over the appropriate internal space when more than one field is involved. It is important to stress that the above expression for the Wilsonian action is *not* exact. However, as we will see in the next subsection, considering an infinitesimal RG transformation allows to derive an *exact* equation which is capable of describing the renormalization group flow of infinitely many couplings.

2.3.2 Wegner-Houghton Equation

The Renormalization Group transformations considered so far deal with an integration over a finite momentum shell $[k, \Lambda[$. Such an integration furnishes an approximate expression for the average action $S_k[\Phi]$, but it does not allow to study the renormalization group flow. In order to address this issue, it is crucial to analyze the variation of the action S_k under an infinitesimal RG transformation. The idea is to start from the action $S_k[\Phi]$ and integrate the Fourier modes belonging to the infinitesimal momentum shell $[k - \delta k, k]$. According to the previous discussion, this procedure gives rise to a new effective action $S_{k-\delta k}[\Phi]$ which is related to $S_k[\Phi]$ by means of

$$\frac{k(S_{k-\delta k}[\Phi]-S_k[\Phi])}{\delta k} = \frac{1}{2}\frac{k}{\delta k}\left\{ \mathrm{STr}'\left[\log(S_k^{(2)}[\Phi])\right] - \mathrm{Tr}'\left[S_k^{(1)}[\Phi](S_k^{(2)}[\Phi])^{-1}S_k^{(1)}[\Phi]\right] + \cdots \right\}. \tag{2.30}$$

Since $\delta k \ll 1$ is assumed, each loop integration yields a volume term $\sim \delta k/k$ [15]. All higher loop contributions included in the "dots" are thus suppressed in the limit $\delta k \to 0$. In this limit, the previous approximate (one loop) equation becomes an exact integro-differential equation for the action functional $S_k[\Phi]$

$$k\partial_k S_k[\Phi] = -\lim_{\delta k \to 0}\frac{k}{\delta k}\left\{\frac{1}{2}\mathrm{STr}'\left[\log(S_k^{(2)}[\Phi])\right] - \frac{1}{2}\mathrm{Tr}'\left[S_k^{(1)}[\Phi](S_k^{(2)}[\Phi])^{-1}S_k^{(1)}[\Phi]\right]\right\}. \tag{2.31}$$

For instance, for a scalar theory with $\mathcal{Z}_2$ symmetry one obtains

$$k\partial_k S_k[\Phi] = -\frac{\Omega_D}{2}\frac{k^D}{(2\pi)^D}\log\left(S_k^{(2)}[\Phi]\right) \tag{2.32}$$

where Ω_D is the volume of the $(D-1)$-dimensional unit sphere. This flow equation, derived by Wegner and Houghton in 1972 [14] and subsequently improved by Polchinski [16], can be considered as the first Exact Renormalization Group Equation (ERGE). The Wegner-Houghton equation describes the flow of the Wilsonian action $S_k[\phi]$ through the entire theory space. It is important to notice that this RG equation does not depend on the UV cutoff Λ, which can also be taken infinite.

Although Eq. (2.31) is formally exact, it is not possible to find a general analytical solution. One possible strategy is that of expanding the effective action $S_k[\Phi]$ in powers of field derivatives [17–19]

$$S_k[\Phi] = \int d^D x \{U_k[\Phi] + \tfrac{1}{2}(\partial_\mu\Phi)^2 Z_k[\Phi] + (\partial_\mu\Phi)^4 Y_k[\Phi] + (\Box\Phi)^2 H_k[\Phi] + \dots\} . \tag{2.33}$$

The simplest approximation one can do is the Local Potential Approximation (LPA). It represents the zeroth-order approximation of the derivative expansion (2.33) and corresponds to the assumption that the field Φ is uniform through spacetime. In fact, if the field is $\Phi = \mathrm{const}$, the effective action $S_k[\Phi]$ reduces to the function $S_k(\Phi) = \Omega\, U_k(\Phi)$, where Ω is the total spacetime volume. In the LPA approximation the Wegner-Houghton Eq. (2.32) assumes the simple form [13]

$$k\,\partial_k U_k(\Phi) = -\frac{k^D}{\Gamma(D/2)(4\pi)^{D/2}}\log\left(k^2 + U_k^{(2)}(\Phi)\right) \tag{2.34}$$

where $U_k(\phi)$ is the natural generalization of the standard effective potential [20]. At this point, the strategy is to expand the potential as

$$U_k(\Phi) = \sum_{n=0}^{N} g_n(k)\, \Phi^n, \qquad (2.35)$$

where N defines the order of the truncation scheme, and then project the flow equation onto the subspace spanned by (2.35). The resulting subset of coupling constants will satisfy a set of ordinary differential equations in the form

$$k\partial_k\, g_n(k) = \beta_n(g_1, \ldots, g_N) \qquad (2.36)$$

where the right-hand-side is completely defined by the *beta functions* β_n. The zeros of these functions furnish the fixed point solutions of the renormalization group flow. In the next section we shall examine the properties of the fixed point solutions, completing the analogy with the case of statistical field theory, and we will discuss the important relation between fixed points and renormalizability.

2.4 Fixed Points and Generalized Renormalizability

The Wegner-Houghton equation captures the renormalization group flow of infinitely many couplings. However, as we have seen, this equation cannot be analytically solved and one has to project the flow into a subspace of the theory space. The resulting set of Eq. (2.36) is analogous to that of Eq. (2.17). By lowering the infrared cutoff k from the ultraviolet regime ($k \to \infty$) to the long-distance limit ($k \to 0$), the *running couplings* $g_n(k)$ draw a trajectory in the theory space. This evolution is again parametrized by the RG time $\tau = \log k$, and all possible "asymptotic states" are determined by the Fixed Point (FP) structure of the renormalization group flow. The stability properties of each fixed point can then be studied by means of the stability matrix $\mathcal{S}_{ij} \equiv \partial_{g_j}\beta_i(\mathbf{g}_*)$.

As we have already seen, a Gaussian Fixed Point (GFP) describes a free theory. Conventionally it corresponds to the origin of the theory space. Notably, as for a trivial fixed point $\xi = 0$, the scaling exponents will be defined by the canonical mass dimensions of the corresponding operators. On the other hand, a Non-Gaussian Fixed Point (NGFP) represents an interacting theory and its critical exponents generally differ from the canonical ones by an anomalous dimension contribution. Just like the case of statistical physics, the scaling operators associated with a particular fixed point can be classified according to their scaling exponents. The relevant coupling constants increase by lowering the energy scale, while the irrelevant couplings become small in the infrared limit. Hence, the relevant observables are the only parameters needed to describe effective field theories and they can be fixed by experiments.

In this discussion it is of central importance to specify the fixed point one refers to. An operator which is irrelevant with respect to one fixed point, may be relevant for another fixed point. The perturbative expansions used in Quantum Field Theory to compute scattering amplitudes are valid when all couplings are small, i.e. near the GFP. Hence the "renormalizable" interactions of perturbative Quantum Field Theory correspond, in the Wilsonian picture, to those operators which are relevant with *respect to the GFP*. Since the perturbation theory is valid in proximity to the GFP, the perturbative notion of renormalizability is related to the existence of such relevant couplings. In fact, for a theory involving just a set of relevant operators, the renormalization group flow approaches the GFP in the ultraviolet limit and the theory is (perturbatively) renormalizable. Conversely, the irrelevant operators drive the renormalization group flow away from the GFP. Therefore, from a perturbative point of view, non-renormalizable operators (irrelevant directions) are associated with irremovable ultraviolet divergences. In their presence the perturbative quantization of the theory leads to several inconsistencies.

The conclusion that the theory is sick cannot be drawn unless the complete fixed point structure of the renormalization group flow is known. The non-perturbative techniques developed to study the Renormalization Group led in fact to a change of perspective. A Quantum Field Theory is fully consistent if and only if its renormalization group flow is well defined for all energy scales [21] . In particular, according to [22], the fundamental requirement is the existence of a UV-attractive fixed point, coming with a finite-dimensional UV critical surface (basin of attraction). Assuming that such a FP exists, all RG trajectories belonging to the critical manifold run towards this fixed point in the ultraviolet limit. On this basis, such an ultraviolet fixed point defines the UV completion of the theory under consideration.

On the basis of these arguments, the Wilsonian Renormalization Group naturally lead to a generalized notion of renormalizability. Even if a theory (RG trajectory) is perturbatively non-renormalizable, it may be renormalizable from a non-perturbative point of view if the renormalization group flow, repelled by the GFP, is attracted to a NGFP. The resulting theory is called *asymptotically safe* because, despite of the presence of perturbatively non-renormalizable operators, the NGFP saves the theory from unphysical divergences. The NGFP makes the theory well defined up to arbitrary large energy scales and perfectly acceptable and consistent. On the other hand, if the ultraviolet fixed point is trivial, this generalized renormalizability criterion reduces to the perturbative one, and the underlying theory becomes *asymptotically free*.

References

1. E.C.G. Stueckelberg, A. Petermann, La renormalisation des constants dans la théorie de quanta. Helv. Phys. Acta **26**, 499–520 (1953). https://doi.org/10.5169/seals-112426. (cit. on p. 17)
2. M. Gell-Mann, F.E. Low, Quantum electrodynamics at small distances. Phys. Rev. **95**, 1300–1312 (1954). https://doi.org/10.1103/PhysRev.95.1300. (cit. on p. 17)
3. L.P. Kadanoff, Scaling laws for Ising models near T(c). Phys. Physique Fizika **2**, 263–272 (1966). https://doi.org/10.1103/PhysicsPhysiqueFizika.2.263. (cit. on p. 17)

4. K.G. Wilson, Renormalization group and critical phenomena. I. Renormalization group and the Kadanoff scaling picture. Phys. Rev. B **4**, 3174–3183 (1971). https://doi.org/10.1103/PhysRevB.4.3174. (cit. on p. 18)

5. K.G. Wilson, Renormalization group and critical phenomena. II. Phase-space cell analysis of critical behavior. Phys. Rev. B **4**, 3184–3205 (1971). https://doi.org/10.1103/PhysRevB.4.3184. (cit. on p. 18)

6. K.G. Wilson, M.E. Fisher, Critical exponents in 3.99 dimensions. Phys. Rev. Lett. **28**, 240–243 (1972). https://doi.org/10.1103/PhysRevLett.28.240. (cit. on p. 18)

7. M.K. Grover, L.P. Kadanoff, F.J. Wegner, Critical exponents for the heisenberg model. Phys. Rev. B **6**, 311–313 (1972). https://doi.org/10.1103/PhysRevB.6.311. (cit. on p. 18)

8. N. Goldenfeld, Lectures on phase transitions and the renormalization group. Front. phys. (Avalon Publishing, 1992). ISBN: 9780201554090 (cit. on pp. 19, 24)

9. L.M. Holmes, L.G. Van Uitert, G.W. Hull, Magnetoelectric effect and critical behavior in the Ising-like antiferromagnet, DyAlO$_3$. Solid State Commun. **9**, 1373–1376 (1971). https://doi.org/10.1016/0038-1098(71)90398-X. (cit. on p. 19)

10. M. Ley-Koo, M.S. Green, Revised and extended scaling for coexisting densities of SF6. Phys. Rev. A 16, 2483–2487 (1977). https://doi.org/10.1103/PhysRevA.16.2483. (cit. on p. 19)

11. K.G. Wilson, The renormalization group—introduction. In: C. Domb, M.S. Green, J.L. Lebowitz (eds.), *Phase Transitions and Critical Phenomena*, vol. 6. (Academic Press, 1976). ISBN: 9780122203060 (cit. on pp. 22, 27)

12. F.J. Wegner, The critical state, general aspects. In: C. Domb, M.S. Green, J.L. Lebowitz (eds.) *Phase Transitions and Critical Phenomena*, Vol. 6. (Academic Press, 1976). ISBN: 9780122203060 (cit. on pp. 25, 26)

13. S.B. Liao, J. Polonyi, Blocking transformation in field theory. Ann. Phys. **222**, 122–156 (Feb. 1993). https://doi.org/10.1006/aphy.1993.1019. (cit. on pp. 27, 28, 30)

14. F.J. Wegner, A. Houghton, renormalization group equation for critical phenomena. Phys. Rev. A **8**, 401–412 (1973). https://doi.org/10.1103/PhysRevA.8.401. (cit. on pp. 28, 30)

15. S.B. Liao, J. Polonyi, Renormalization group and universality. Phys. Rev. D **51**, 4474–4493 (1995). https://doi.org/10.1103/PhysRevD.51.4474.eprint:hep-th/9403111. (cit. on pp. 28, 29)

16. J. Polchinski, Renormalization and effective lagrangians. Nuclear Phys. B **231**, 269–295 (1984). https://doi.org/10.1016/0550-3213(84)90287-6. (cit. on p. 30)

17. T.R. Morris, Derivative expansion of the exact renormalization group. Phys. Lett. B **329**, 241–248 (1994). https://doi.org/10.1016/0370-2693(94)90767-6.eprint:hep-ph/9403340. (cit. on p. 30)

18. T.R. Morris, On truncations of the exact renormalization group. Phys. Lett. B **334**, 355–362 (1994). https://doi.org/10.1016/0370-2693(94)90700-5.eprint:hep-th/9405190. (cit. on p. 30)

19. T.R. Morris, The derivative expansion of the renormalization group. Nuclear Phys. B Proc. Suppl. **42**, 811–813 (1995). https://doi.org/10.1016/0920-5632(95)00389-Q.eprint:hep-lat/9411053. (cit. on p. 30)

20. L.H. Ryder, *Quantum Field Theory*. (Cambridge University Press, 1985). ISBN: 9780521237642 (cit. on p. 30)

21. S. Weinberg, Critical phenomena for field theorists, in *Proceedings 14th International School of Subnuclear Physics. Erice* (1976), p. 1. https://doi.org/10.1007/978-1-4684-0931-4_1(cit. on p. 32)

22. K.G. Wilson, J. Kogut, The renormalization group and the ϵ expansion. Phys. Rep. **12**, 75–199 (1974). https://doi.org/10.1016/0370-1573(74)90023-4. (cit. on p. 32)

Chapter 3
Functional Renormalization and Asymptotically Safe Gravity

Einstein's theory of General Relativity provides a successful framework for the description of the gravitational interaction, but it breaks down in those regimes where the spacetime curvature diverges, and the spacetime becomes singular. In such ultra-Planckian regimes we expect Quantum Gravity effects to become important. However, a fully consistent theory of Quantum Gravity is still lacking and this issue represents one of the major unsolved problems in theoretical physics. The main problem arising in the construction of a quantum theory for the gravitational interaction is that General Relativity turns out to be (perturbatively) non-renormalizable. Most of the theories proposed so far are based on radical rethinking of quantum theory and require the introduction of new physics. An exception is the Asymptotic Safety scenario for Quantum Gravity, which is based on pure Quantum Field Theory. It builds on the generalized notion of renormalizability naturally arising from the Wilsonian Renormalization Group.

The Wilsonian formulation of the Renormalization Group has led to a deep understanding of the meaning of renormalization in Quantum Field Theory. By means of a generalization of the Kadanoff blocking transformation and the functional methods developed in the context of Quantum Field Theory, Wegner and Houghton derived an exact equation [1] capable of describing the evolution of the renormalization group flow through the theory space. The resulting (non-perturbative) FRG methods allow to investigate the strongly-interacting regimes characterizing the physics far away from the GFP. In this framework it is possible to study the complete fixed point structure of the renormalization group flow. In particular the method allows to search for non-trivial fixed points inaccessible by the standard perturbative methods. In particular, as proposed by Wilson and Kogut [2] in 1974, the high-energy completion of a Quantum Field Theory can be defined if there exist a fixed point which attracts the renormalization group flow in the ultraviolet (UV) limit. As we have seen in the previous chapter, this observation leads to a generalized notion of renormalizability based on the existence of UV-attractive non-trivial fixed points. Perturbatively non-renormalizable quantum field theories can thus be renormalizable from a

© Springer Nature Switzerland AG 2018
A. B. Platania, *Asymptotically Safe Gravity*, Springer Theses,
https://doi.org/10.1007/978-3-319-98794-1_3

non-perturbative point of view. From this standpoint, as first proposed by Weinberg in [3], gravity may be a perfectly renormalizable Quantum Field Theory and the existence of a NGFP may provide the UV-completion of the gravitational interaction. This framework is commonly referred to as Asymptotically Safe scenario for Quantum Gravity.

Due to the difficulties related to the study of gauge theories within the FRG approach, the progresses along this line of research slowed down for several years. In the 90's, new functional methods based on the concept of Effective Average Action (EAA) were developed and a Functional Renormalization Group Equation (FRGE) for scalar field theories, known as Wetterich equation, was derived [4, 5]. The subsequent generalization of the EAA formalism to gauge theories [6] renewed the interest in the possibility of finding a NGFP underlying the renormalizability of the gravitational interaction. In 1996, M. Reuter derived an exact equation describing the renormalization group flow of the gravitational interaction [7]. This result represented the starting point for a systematic investigation of the Asymptotic Safety scenario for Quantum Gravity.

Although the Asymptotic Safety conjecture is still unproved, there is a number of evidences supporting its validity. In particular, for the case where the gravitational degrees of freedom are encoded in fluctuations of the (Euclidean) spacetime metric, defining the so-called *metric approach to Asymptotic Safety*, the existence of a suitable NGFP has been shown in a vast number of approximations [8, 9, 11–35].

In this chapter we shall review the basic FRG techniques developed during the 90's. Subsequently we will apply the EAA formalism to the case of Quantum Gravity, summarizing the fundamental results obtained by Reuter in [7] and describing the features of the corresponding phase diagram, first constructed in [10].

3.1 The Wetterich Equation for Scalars

The Wegner-Houghton equation (2.31) is a formally exact equation and its derivation requires the introduction of an infrared cutoff k which allows to define the Wilsonian average action. In the 90's new functional techniques based on the concept of the Effective Average Action (EAA) were introduced and resulted in the derivation of another ERGE, known as Wetterich equation [4, 5].

In contrast to the Wegner-Houghton equation, the Wetterich equation makes use of a smooth cutoff function and its derivation is very similar to that of the standard effective action in Quantum Field Theory. In the Wetterich formalism the dependence of the effective action on the infrared scale k is encoded in a smooth cutoff function $\Delta_k S$ defined as

$$\Delta_k S[\phi] \equiv \frac{1}{2} \int \phi^*(-q) \mathcal{R}_k(q^2) \phi(q) \, \frac{d^D q}{(2\pi)^D}. \qquad (3.1)$$

The cutoff operator $\mathcal{R}_k(q^2)$ is a matrix valued function and plays the role of an effective mass term suppressing all fluctuations with momenta $q < k$. Explicitly $\mathcal{R}_k(q^2)$ can be rewritten as $\mathcal{R}_k(q^2) \equiv Z_k R_k(q^2)$, where the scalar function $R_k(q^2)$ smoothly interpolates between the two regimes

$$R_k(q^2) = \begin{cases} 0 & q \gg k \\ k^2 & q \ll k \end{cases} \tag{3.2}$$

so that the slow Fourier modes are suppressed by the dynamical mass k^2, and the functional integration is automatically restricted to the only fast Fourier modes.

The cutoff function $\Delta_k S$ is explicitly introduced in the functional integral, thus defining a scale-dependent generating functional

$$\mathcal{Z}_k[J] \equiv e^{W_k[J]} = \int \mathcal{D}[\phi] \, e^{-S_\Lambda[\phi] - \Delta_k S[\phi] + J \cdot \phi} \tag{3.3}$$

with $J(x)$ an external source field. The scale-dependent functional $W_k[J]$ allows to define the effective average field as

$$\Phi[x; J] \equiv \langle \phi \rangle_k = \frac{\delta W_k[J]}{\delta J(x)}. \tag{3.4}$$

The source field can thus be thought as a functional of the field Φ, $J = J[x; \Phi]$. By recalling how the standard effective action is obtained in Quantum Field Theory [36], it is natural to define the EAA $\Gamma_k[\Phi]$ by means of the Legendre transform of $W_k[J]$. Because of the explicit introduction of the cutoff function $\Delta_k S$, the Legendre transform of $W_k[J]$ is the functional $\widetilde{\Gamma}_k[\phi] = \Gamma_k[\phi] + \Delta_k S[\phi]$,[1] with the EAA $\Gamma_k[\phi]$ defined as

$$\Gamma_k[\Phi] \equiv (-W_k[J] + J[\Phi] \cdot \Phi) - \tfrac{1}{2} \Phi \cdot \mathcal{R}_k \cdot \Phi. \tag{3.5}$$

In particular the external source $J[x; \Phi]$ can be written as the following functional derivative

$$J[x; \Phi] \equiv \frac{\delta(\Gamma_k[\Phi] + \Delta_k S[\Phi])}{\delta \Phi}. \tag{3.6}$$

At this point the flow equation for the action functional $\Gamma_k[\Phi]$ can be obtained by differentiating Eq. (3.5) with respect to the infrared scale k

$$\begin{aligned} \partial_k \Gamma_k[\Phi] &= -\partial_k W_k[J] - \tfrac{1}{2} \Phi \cdot (\partial_k \mathcal{R}_k) \cdot \Phi \\ &= +\tfrac{1}{2} \left(\frac{\delta W_k[J]}{\delta J} \cdot \frac{\partial \mathcal{R}_k}{\partial k} \cdot \frac{\delta W_k[J]}{\delta J} + \mathrm{Tr}\left[\left(\frac{\delta J[\Phi]}{\delta \Phi} \right)^{-1} \frac{\partial \mathcal{R}_k}{\partial k} \right] \right) - \tfrac{1}{2} \Phi \cdot \frac{\partial \mathcal{R}_k}{\partial k} \cdot \Phi. \end{aligned} \tag{3.7}$$

---

[1] Since the action $\Gamma_k[\Phi]$ is not the Legendre transform of $W_k[J]$, it is not restricted to be a convex functional. The convexity properties of the effective action can only be recovered in the limit $k \to 0$.

Combining the above equation with (3.6) we conclude that

$$k\partial_k \Gamma_k[\Phi] = \tfrac{1}{2} \operatorname{Tr} \left[(\Gamma_k^{(2)} + \mathcal{R}_k)^{-1} \, k\partial_k \mathcal{R}_k \right].$$ (3.8)

Here $\Gamma_k^{(2)}$ is the second functional derivative of $\Gamma_k[\Phi]$ with respect to the field Φ. Analogously to the Wegner-Houghton equation, the Wetterich equation (3.8) describes the renormalization group flow of a scale-dependent average action. The effective action $\Gamma_k[\Phi]$ is supposed to smoothly interpolate between the ultraviolet limit and the low-energy regime. Compared to the Wegner-Houghton approach, the EAA has the advantage of providing a direct relation to the connected Green's functions. In particular, by noting that $\frac{\delta \Phi[J]}{\delta J} \equiv W_k^{(2)}$ and $\frac{\delta J[\Phi]}{\delta \Phi} \equiv \Gamma_k^{(2)} + \mathcal{R}_k$, we can deduce the following identity

$$W_k^{(2)} (\Gamma_k^{(2)} + \mathcal{R}_k) \equiv \mathbb{1}$$ (3.9)

Therefore, the functional $(\Gamma_k^{(2)} + \mathcal{R}_k)^{-1}$ represents an effective (modified) propagator at the scale k. As already mentioned, even though the FRG techniques allow a non-perturbative description of the renormalization group flow, the exact Eqs. (2.31) and (3.8) cannot be solved exactly. On the other hand, once an ansatz for the EAA has been chosen, the beta functions determine an approximate renormalization group flow living in the sub-theory space associated with the initial ansatz. This strategy allows to investigate the fixed point structure of the theory.

3.2 The Wetterich Equation for Gauge Theories

In this section we will extend the FRG techniques to gauge theories and in particular to the case of gravity.

3.2.1 Gauge Fixing and Ghosts

In the path integral quantization the key object is the generating functional $\mathcal{Z}[J]$. This functional allows to compute the Green's functions of any Quantum Field Theory and its expression is determined by a functional integral over *all* field configurations. However, if the theory has some internal (gauge) symmetry, the functional integral would extend over an infinite number of equivalent configurations and thus $\mathcal{Z}[J]$ would be ill defined. More precisely, the redundant field configurations are those obtained under gauge transformations and must be removed from the functional integral by fixing the gauge. A method to implement a gauge-fixing within the path-integral approach was developed by Faddeev and Popov in 1967 [37].

Let us consider a gauge field $A(x)$, and let $G_\mu(A)$ be a function such that $G_\mu(A) = 0$ defines a gauge-fixing condition. According to the Faddeev-Popov method, the gauge-fixing condition can be inserted in the functional integral by means of the identity

$$\int \mathcal{D}[\tilde{\alpha}_\nu] \, \delta[G_\mu(A^{\tilde{\alpha}_\nu})] \det\left(\frac{\delta G(A^{\tilde{\alpha}})}{\delta\tilde{\alpha}}\right) = \mathbb{1} \tag{3.10}$$

where $\tilde{\alpha}(x)$ denotes the set of auxiliary functions $\tilde{\alpha}_\nu(x)$ such that $A^{\tilde{\alpha}_\nu}(x)$ describes all configurations gauge-equivalent to $A(x)$, and $G(A)$ briefly denotes the set of conditions $\{G_\mu\}$. Requiring the functional $G(A^{\tilde{\alpha}})$ to be linear, the Faddeev-Popov determinant can be isolated and treated as a constant. Therefore, the integration over $\tilde{\alpha}_\nu(x)$ reduces to a multiplicative factor $\mathcal{N}$ defining the "volume" of the gauge group [38]. The functional integral thus reads

$$\mathcal{Z}[J_A] = \mathcal{N} \det\left(\frac{\delta G(A^{\tilde{\alpha}})}{\delta\tilde{\alpha}}\right) \int \mathcal{D}[A] \, \delta[G_\mu(A)] \, e^{-S[A]+J_A \cdot A}. \tag{3.11}$$

Without loss of generality we can assume the gauge-condition to be in the form $G_\mu(A) \equiv F_\mu[A] - \omega_\mu$. Since $\omega_\mu(x)$ is an arbitrary gauge invariant function, we can rewrite Eq. (3.11) as follows

$$\mathcal{Z}[J_A] = \mathcal{N} \det\left(\frac{\delta F(A^{\tilde{\alpha}})}{\delta\tilde{\alpha}}\right) \int \mathcal{D}[A] \, e^{-S[A]+J_A \cdot A} \int \mathcal{D}[\omega_\mu] \, \delta[F_\mu[A] - \omega_\mu] \, e^{-\int \frac{\omega_\mu \omega^\mu}{2\alpha} d^D x}$$

$$= \mathcal{N} \det\left(\frac{\delta F(A^{\tilde{\alpha}})}{\delta\tilde{\alpha}}\right) \int \mathcal{D}[A] \, e^{-S[A]+J_A \cdot A} \, e^{-\int (F_\mu F^\mu/2\alpha) d^D x}. \tag{3.12}$$

The resulting exponential can be regarded as a *gauge-fixing action*, and α is the corresponding gauge parameter. Now it only remains to study the Faddeev-Popov determinant. This determinant is taken into account by introducing a new set of anti-commuting fields $\{\bar{C}_\mu, C_\mu\}$ and "exponentiating" the determinant as follows

$$\det \mathcal{M} \equiv \det\left(\frac{\delta F(A^{\tilde{\alpha}})}{\delta\tilde{\alpha}}\right) = \int \mathcal{D}[\bar{C}_\mu] \mathcal{D}[C_\mu] \, e^{-\int (\bar{C}^\mu \mathcal{M}_{\mu\nu} C^\nu) d^D x} \tag{3.13}$$

with $\mathcal{M}_{\mu\nu} = \frac{\delta F_\mu}{\delta\tilde{\alpha}^\nu}$. The set of fields introduced take the name of ghosts, and their introduction in the field integral concludes the Faddeev-Popov procedure. This method allows to remove all gauge-equivalent field configurations and lead to the following generating functional

$$\mathcal{Z}[J] = \mathcal{N} \int \mathcal{D}[\chi] \exp\left\{-S[\chi] - \int \frac{F_\mu F^\mu}{2\alpha} d^D x - \int (\bar{C}^\mu \mathcal{M}_{\mu\nu} C^\nu) d^D x + J \cdot \chi\right\} \tag{3.14}$$

where χ denotes the set $\chi \equiv \{A, \bar{C}_\mu, C_\mu\}$ and J stands for the corresponding set of sources $J \equiv \{J_A, \sigma_\mu, \bar{\sigma}_\mu\}$. The functional (3.14) can now be taken as starting point for the generalization of the EAA formalism to gauge theories.

3.2.2 Background Field Method

With the purpose of defining a scale-dependent action for the gravitational field, one of the most important requirements is the principle of background independence. The diffeomorphism invariance is in fact the gauge symmetry underlying the theory of General Relativity and therefore it must be recovered at least in the infrared limit, $k \to 0$. The problem lies in the path integral quantization procedure. In fact, the gauge-fixing and ghost terms generally break this symmetry and, in the EAA formalism, the same holds for the cutoff action $\Delta_k S$. Furthermore, when considering General Relativity, the metric defining the scalar products (in particular the infrared scale k) is the same dynamical field over which the functional integral is extended over. The Background Field Method (BFM) provides a tool to implement the principle of background independence. This method is essential to define the EAA for the gravitational interaction while preserving the diffeomorphism invariance.

The background field formalism was initially introduced in the context of Quantum Field Theory to quantize gauge theories without losing the gauge invariance [39, 40]. The key idea is to split the physical field into a classical background $\bar{\phi}$ and the fluctuation field $\hat{\phi}$

$$\phi = \bar{\phi} + \hat{\phi}. \tag{3.15}$$

It is important to stress that the fluctuations described by $\hat{\phi}$ are not required to be "small". As the background is fixed, the partition function can then be defined as the following function of the background field

$$\mathcal{Z}[J; \bar{\phi}] \equiv e^{W[J; \bar{\phi}]} \int \mathcal{D}[\hat{\phi}] \, e^{-S_\Lambda[\bar{\phi} + \hat{\phi}] + J \cdot \hat{\phi}} \equiv \mathcal{Z}[J; 0] e^{-J \cdot \bar{\phi}}. \tag{3.16}$$

The functional $\mathcal{Z}[J; \bar{\phi}]$ allows to derive the expectation value of the field $\hat{\phi}$, $\hat{\Phi} = \langle \hat{\phi} \rangle$, and the corresponding effective action in the standard way. The generating functional $W[J; \bar{\phi}]$ is related to $W[J; 0]$ through the relation

$$W[J; \bar{\phi}] = W[J; 0] - J \cdot \bar{\phi} \tag{3.17}$$

and thus the classical field $\Phi = \langle \phi \rangle$ can be decomposed as $\Phi = \bar{\Phi} + \hat{\Phi}$. At this point the remaining task is that of determining the relation between the effective action functionals with and without the background field

$$\Gamma[\hat{\Phi}; \bar{\Phi}] = -W[J; \bar{\Phi}] + J \cdot \hat{\Phi} = -W[J] + J \cdot (\bar{\Phi} + \hat{\Phi}) \equiv \Gamma[\bar{\Phi} + \hat{\Phi}]. \tag{3.18}$$

This relation means that the effective action for the fluctuation field $\hat{\Phi}$ in the presence of a background $\bar{\Phi}$ is equivalent to the usual effective action $\Gamma[\Phi]$ if the physical field Φ is decomposed as $\Phi = \bar{\Phi} + \hat{\Phi}$. The crucial observation is that $\Gamma[0; \bar{\Phi}] = \Gamma[\bar{\Phi}]$. The gauge-fixing can thus be constructed in a way that guarantees the gauge invariance of the standard effective action. In fact, the linear split (3.15) allows to realize the transformation $\Phi \rightarrow \Phi + \delta\Phi$ of the original gauge group as the overlap of a background gauge transformation and a "quantum" gauge transformation. While the former acts on both the background and fluctuation fields, $\Phi \rightarrow \Phi + \bar{\delta}\Phi$, the latter leaves the background invariant, $\Phi \rightarrow \Phi + \hat{\delta}\hat{\Phi}$. Hence, the quantum gauge invariance corresponds to the symmetry respect to which the gauge-fixing has to be added. Only the quantum gauge symmetry is thus explicitly broken by the gauge-fixing and regulator terms, while the whole action remains invariant under background gauge transformations. The latter property guarantees that the standard effective action $\Gamma[0; \bar{\Phi}] = \Gamma[\bar{\Phi}]$ preserves the gauge symmetry.

According to the above discussion, the background field formalism is perfectly suited to study the quantization of the gravitational field within the FRG approach. The fluctuations, central object of the functional renormalization, are integrated out leaving the diffeomorphism invariance of the effective action unaffected. In particular, the metric degree of freedom can be decomposed through the linear split

$$g_{\mu\nu} = \bar{g}_{\mu\nu} + h_{\mu\nu} \tag{3.19}$$

and the generating functional reads

$$\mathcal{Z}_k[J; \bar{\chi}] = \int \mathcal{D}[\hat{\chi}] \, e^{-S^{\text{grav}} - S^{\text{gf}} - S^{\text{ghost}} - \Delta_k S + J \cdot \hat{\chi}} \tag{3.20}$$

where χ is a set of fields which includes the metric $g_{\mu\nu}$ and the Faddeev-Popov ghosts, and J is the corresponding set of sources. The cutoff action $\Delta_k S$ has to be constructed through the BFM and is given by

$$\Delta_k S = \frac{1}{2} \int \sqrt{\bar{g}} \left\{ \hat{\chi} \mathcal{R}_k[\bar{\chi}] \hat{\chi} \right\} d^D x. \tag{3.21}$$

Finally, the action functional Γ_k can be decomposed into the physical action, and the gauge-fixing and ghost contributions. For pure gravity Γ_k can thus be written as

$$\Gamma_k[\hat{\chi}; \bar{\chi}] \equiv \Gamma_k^{\text{grav}}[h; \bar{g}] + \Gamma_k^{\text{gf}}[h; \bar{g}] + \Gamma_k^{\text{ghost}}[h, C, \bar{C}; \bar{g}] \tag{3.22}$$

with

$$\Gamma_k^{\text{gf}}[h; \bar{g}] = \frac{1}{2\alpha} \int \sqrt{\bar{g}} \left\{ \bar{g}^{\mu\nu} F_\mu F_\nu \right\} d^D x \ , \tag{3.23a}$$

$$\Gamma_k^{\text{ghost}}[C, \bar{C}, h; \bar{g}] = \int \sqrt{\bar{g}} \left\{ \bar{C}_\mu \mathcal{M}_\nu^\mu C^\nu \right\} d^D x, \tag{3.23b}$$

and where the functional $F_\mu[h; \bar{g}]$ fixes the gauge through $F_\mu[h; \bar{g}] = 0$. At last, for a given truncation scheme, the left-hand-side of the Wetterich equation can be determined by differentiating the action functional (3.22) with respect to the Renormalization Group time $\tau = \log k$. This variation basically involves the derivatives of the couplings with respect to k, and the comparison with the right-hand-side allow to read off the corresponding beta functions.

3.2.3   Cutoff Function and Heat-Kernel Techniques

Once an ansatz for the EAA has been chosen, and the gauge fixing and ghosts terms have been added, the left-hand-side of the flow equation is completely determined. The next step of the procedure is to compute the right-hand-side. Typically this can be done by using the off-diagonal heat-kernel technique (see [41] and the Appendix of [15] for details). In this section we will show how this method works for a physical system described by a generic set of fields $\chi(x)$, which can include scalar, vector and tensor fields, as well as the Faddeev-Popov ghosts.

The basic ingredient in the right-hand-side is the cutoff function $\mathcal{R}_k$ which has the role of regularizing the inverse propagator $\Gamma_k^{(2)}$ (Hessian). Following the notation in [41], the latter differential operator can be written as

$$\Gamma_k^{(2)} = \mathbb{K} + \mathbb{D} + \mathbb{M}. \tag{3.24}$$

The operators $\mathbb{K}$, $\mathbb{D}$ and $\mathbb{M}$ denote the kinetic, uncontracted derivatives and background interaction terms respectively. In order to make the computation easier, and use the standard heat-kernel techniques, it is useful to "eliminate" the $\mathbb{D}$-terms. The standard procedure consists of using a Transverse-Traceless (TT), or York, decomposition of the fluctuation fields. For instance, vector and tensor fields can be decomposed into their transverse and longitudinal parts as follows

$$\hat{A}_\mu = \hat{A}_\mu^T + \bar{D}_\mu B \ , \tag{3.25a}$$

$$h_{\mu\nu} = h_{\mu\nu}^{TT} + \bar{D}_\mu \xi_\nu + \bar{D}_\nu \xi_\mu - \tfrac{2}{D} \bar{g}_{\mu\nu} \bar{D}^\gamma \xi_\gamma + \tfrac{1}{D} \bar{g}_{\mu\nu} h. \tag{3.25b}$$

Here $\bar{D}_\mu$ denotes the background covariant derivative, $\hat{A}_\mu^T$ is a transverse vector field ($\bar{D}^\mu \hat{A}_\mu = 0$), $h_{\mu\nu}^{TT}$ is a transverse-traceless tensor field ($\bar{D}^\mu h_{\mu\nu}^{TT} = 0$ and $\bar{g}^{\mu\nu} h_{\mu\nu}^{TT} = 0$), and $h \equiv \bar{g}^{\mu\nu} h_{\mu\nu}$ stands for the trace of $h_{\mu\nu}$. In particular, the vectors ξ_μ can be further decomposed according to (3.25a). When applying the TT-decomposition defined above to the set of fluctuation fields $\hat{\chi}$ appearing in the second variation of the action, $\delta^2 \Gamma_k \equiv \tfrac{1}{2} \hat{\chi} \cdot \Gamma_k^{(2)} \cdot \hat{\chi}$, all $\mathbb{D}$-terms reduce to contributions to $\mathbb{K}$ and $\mathbb{M}$.

The differential operator $\Gamma_k^{(2)}$ is a block matrix given by the sum of the kinetic $\mathbb{K}$ and interaction $\mathbb{M}$ operators. In particular the operator $\mathbb{K}$ has the structure of a block-diagonal matrix, whose non-zero blocks $\mathbb{K}_{ii}^{(s)}$ are defined on the subspaces spanned by the fields $\hat{\chi}_i$ having s spacetime indices. The blocks $\mathbb{K}_{ij}^{(s)}$ thus read

$$\mathbb{K}_{ij}^{(s)} \propto \delta_{ij} \, \square_s \, \hat{\mathbb{1}}_i \tag{3.26}$$

where $\square_s \equiv -\bar{g}^{\mu\nu} \bar{D}_\mu \bar{D}_\nu$ is the D-dimensional Laplacian acting on fields having s indices, and $\hat{\mathbb{1}}_i$ stands for the identity $\hat{\mathbb{1}}_{d_i \times d_i}$ in the d_i-dimensional internal subspace spanned by the field $\hat{\chi}_i$. The interaction operator $\mathbb{M}$ is usually much more complicated, and can be decomposed in the following way

$$\mathbb{M} = \hat{E} + \hat{Q}_k . \tag{3.27}$$

Here both $\hat{E}$ and $\hat{Q}_k$ contain interaction terms, but the explicit dependence on the running couplings is now enclosed in the scale-dependent operator $\hat{Q}_k$. The most general block element of $\Gamma_k^{(2)}$ can thus be written as

$$\Gamma_k^{(2)}\Big|_{\hat{\chi}_l\hat{\chi}_j} = (\hat{L}_k)_{li} \left(\delta_{ij} \, \Box_{s_i} \, \hat{\mathbb{1}}_i + (\hat{E} + \hat{Q}_k)_{ij} \right) \tag{3.28}$$

where a diagonal scale-dependent operator $(\hat{L}_k)_{ij} \equiv \delta_{ij} \, \gamma_i(k) \, \hat{\mathbb{1}}_i$ has been introduced.

At this point it is important to specify the cutoff scheme, namely how to implement the regularization of $\Gamma_k^{(2)}$ through the operator $\mathcal{R}_k$. As the latter represents a generalized mass-term suppressing the modes with $q < k$, the most natural way to regularize the Hessian is that of replacing

$$\Box_s \mapsto P_k(\Box_s) \equiv \Box_s + R_k(\Box_s) \tag{3.29}$$

where the scalar function $R_k(\Box_s)$ has to be chosen according to the requirement (3.2). The regularization scheme introduced through Eq. (3.29) defines the so-called Type I cutoff $\mathcal{R}_k|_{ij} \equiv \delta_{ij} \, \gamma_i(k) \, R_k(\Box_{s_i})$ [15]. Other cutoff choices are however possible and their implementation make use of Eq. (3.29) with the Laplacian $\Box_s$ replaced by a generic differential operator Δ which includes the interaction terms. Specifically, $\Delta = \Box_s + \hat{E}$ implements the Type II regulator, while the scale-dependent choice $\Delta = \Box_s + \hat{E} + \hat{Q}_k$ corresponds to the Type III regulator, which is also referred to as "spectrally adjusted" cutoff because of its dependence on the running couplings [15]. By using the Type I cutoff, the modified inverse propagator can be written as

$$\tilde{\Gamma}_k^{(2)}\Big|_{\hat{\chi}_l\hat{\chi}_j} = (\Gamma_k^{(2)} + \mathcal{R}_k)\Big|_{\hat{\chi}_l\hat{\chi}_j} = (\hat{L}_k)_{li} \left(P_k(\Box_{s_i}) \, \delta_{ij} \, \hat{\mathbb{1}}_i + (\hat{E} + \hat{Q}_k)_{ij} \right) \tag{3.30}$$

while the numerator in the right-hand-side of the flow equation reads

$$k\partial_k \mathcal{R}_k(\Box_{s_i})\Big|_{\hat{\chi}_l\hat{\chi}_j} = (\hat{L}_k)_{li} \left(k\partial_k R_k + R_k \, k\partial_k \log[\gamma_j(k)] \right) \delta_{ij}. \tag{3.31}$$

At this point we are left with the task of inverting the regularized inverse propagator and evaluate the operator traces by means of the off-diagonal heat-kernel technique. For this purpose it is useful to specialize the discussion to the case of the gravitational field. In this regard, we consider for the gravitational action the following polynomial truncation

$$\Gamma_k^{\mathrm{grav}}[g_{\mu\nu}] = \int \sqrt{g} \left(\sum_{m=0}^{M} g_m(k) R^m \right) d^D x \tag{3.32}$$

where M defines the order of the truncation. As the initial assumption on the physical action is a polynomial approximation, the computation of the beta functions for the running couplings requires a comparison between expansion coefficients in the left and right hand-sides of the Wetterich equation. Therefore, the operator to be traced needs to be developed in powers of the expansion parameter (for instance the curvature R). At this stage it is convenient to decompose

$$\widetilde{\Gamma}_k^{(2)} = \hat{L}_k \left(\widetilde{\mathcal{P}}(\Box_s) + \widetilde{\mathcal{V}}(\bar{R}) \right) \tag{3.33}$$

where the scale-dependent matrix $\widetilde{\mathcal{P}}$ collects all kinetic and constant terms, while the curvature-dependent operators (evaluated on the background) are enclosed in the interaction matrix $\widetilde{\mathcal{V}}$. Hence, putting $\widetilde{\mathcal{N}}(\Box_s) \equiv \hat{L}_k^{-1} k \partial_k \mathcal{R}_k$, the operator in the right-hand-side of the Wetterich equation can be expanded as follows

$$\left(\widetilde{\Gamma}_k^{(2)} \right)^{-1} (k \partial_k \mathcal{R}_k) = \left(\widetilde{\mathcal{P}}^{-1} - \widetilde{\mathcal{P}}^{-1} \widetilde{\mathcal{V}} \widetilde{\mathcal{P}}^{-1} + \widetilde{\mathcal{P}}^{-1} \widetilde{\mathcal{V}} \widetilde{\mathcal{P}}^{-1} \widetilde{\mathcal{V}} \widetilde{\mathcal{P}}^{-1} + \cdots \right) \widetilde{\mathcal{N}}. \tag{3.34}$$

The only point remaining concerns the evaluation of the operator traces, for which the covariant heat-kernel methods are needed [41]. The trace of a general function $W(\Delta)$ of the operator Δ can be written as

$$\text{Tr}\,[W(\Delta)] = \int_0^\infty \left\{ \text{Tr}[K(s)] \right\} \widetilde{W}(s)\, ds \tag{3.35}$$

where the function $\widetilde{W}(s)$ is the inverse Laplace transform of $W(s)$, and $K(s) \equiv e^{-s\Delta}$ is the *heat kernel* of the operator Δ. The trace in the above expression can be developed through the "early-time expansion"

$$\text{Tr}[K(s)] \equiv \sum_i e^{-s\lambda_i} = \int \sqrt{g} \left\{ \frac{1}{(4\pi s)^{D/2}} \sum_{n=0}^\infty \text{Tr}[b_{2n}(\Delta)]\, s^n \right\} d^D x. \tag{3.36}$$

Here λ_i are the eigenvalues of Δ and b_{2n} are the Seeley-Gilkey coefficients [42] of the heat-kernel expansion. In particular each coefficient b_m contains m derivatives and for a differential operator $\Delta = \Box_s + \hat{E}$ the first two coefficients are $b_0 = \hat{\mathbb{1}}$ and $b_2 = \frac{R}{6}\hat{\mathbb{1}} - \hat{E}$. Combining (3.35) and (3.36) finally gives

$$\text{Tr}\,[W(\Delta)] = \frac{1}{(4\pi)^{D/2}} \int \sqrt{g} \left\{ \sum_{n=0}^\infty Q_{\frac{D-2n}{2}}[W]\, \text{Tr}[b_{2n}(\Delta)] \right\} d^D x \tag{3.37}$$

where the functionals Q_n are defined as the Mellin transforms of $W(s)$

$$Q_n[W] = \int_0^\infty \widetilde{W}(s)\, s^{-n}\, ds \equiv \frac{1}{\Gamma(n)} \int_0^\infty W(z)\, z^{n-1}\, dz \tag{3.38}$$

for $n \in \mathbb{N}_0$, while $Q_0[W] = W(0)$ for $n = 0$.

Finally, the formula (3.37) can be used to evaluate the trace in the right-hand-side of the flow equation. Considering a truncation scheme in the form (3.32), involving up to M powers in the curvature scalar, the trace in the right-hand-side can accordingly be approximated by

$$\text{Tr}\left[\frac{k\partial_k \mathcal{R}_k}{\tilde{\Gamma}_k^{(2)}}\right] \simeq \frac{1}{(4\pi)^{D/2}} \sum_{m=0}^{M} \sum_{n=0}^{M-m} (-1)^m \int \sqrt{\bar{g}} \left\{ Q_{\frac{D-2n}{2}} \left[W_{(m)}(z)\right] \text{Tr}[b_{2n}] \right\} v_m \bar{R}^m d^D x$$

(3.39)

where the heat-kernel coefficients are functions of the box operator, $b_{2n} = b_{2n}(\Box_s)$ [42], the coefficients v_m depends on the structure of $\widetilde{\mathcal{V}}(\bar{R})$ and the family of functions $W_{(m)}(z)$ is defined as follows

$$W_{(m)}(z) = \frac{\tilde{N}(z)}{[\tilde{\mathcal{P}}(z)]^{m+1}}.$$

(3.40)

Following the procedures described in this section, we are now ready to analyze the renormalization group flow for the gravitational field in the Einstein-Hilbert truncation.

3.3 Asymptotic Safety in Quantum Einstein Gravity

This section is devoted to the study of the renormalization group flow of pure Quantum Gravity in the Einstein-Hilbert truncation. This ansatz approximates the gravitational part of the EAA by

$$\Gamma_k^{\text{grav}}[g] = \frac{1}{16\pi G_k} \int \sqrt{g} \{-R(g) + 2\Lambda_k\} d^D x$$

(3.41)

where the Newton's coupling and cosmological constant depend on the RG scale k. The left-hand-side of the Wetterich equation can thus be computed by differentiating the action functional (3.41) with respect to the RG time $\tau = \log k$.

The total average action defined in Eq. (3.22) has to satisfy the Wetterich equation

$$k\partial_k \Gamma_k[h; \bar{g}] = \frac{1}{2} \text{STr} \left[\frac{k\partial_k \mathcal{R}_k}{\Gamma_k^{(2)} + \mathcal{R}_k}\right]$$

(3.42)

where $\Gamma_k^{(2)}$ denotes the second functional derivative of the effective action $\Gamma_k[h; \bar{g}]$ with respect to the fluctuation fields

$$\Gamma_k^{(2)}\Big|_{ij} = \epsilon \frac{1}{\sqrt{\bar{g}}} \frac{1}{\sqrt{\bar{g}}} \frac{\delta^2 \Gamma_k}{\delta \hat{\chi}_i \, \delta \hat{\chi}_j}$$

(3.43)

with $\epsilon = 1$ for bosons and $\epsilon = -1$ for fermion fields. In order to derive the beta functions and determine the renormalization group flow in the Einstein-Hilbert subspace, we introduce the dimensionless couplings

$$g_k \equiv k^{D-2} G_k \ , \qquad \lambda_k \equiv k^{-2} \Lambda_k \qquad (3.44)$$

and the anomalous dimension of the Newton's constant, defined as

$$\eta \equiv (G_k)^{-1} k \partial_k G_k = (g_k)^{-1} k \partial_k g_k - (D - 2). \qquad (3.45)$$

Following the notation introduced in the previous section, the scale dependent function $\gamma_i(k)$ is given by

$$\gamma_i(k) = \left(\frac{1}{16\pi G_k} \right)^{\alpha_i} \qquad (3.46)$$

where $\alpha_i = 0, 1$ depending on whether the matrix element arises from the gravitational or ghost sector. Therefore the operators $\widetilde{\mathcal{P}}$ and $\widetilde{\mathcal{N}}$ assume the simple form

$$\widetilde{\mathcal{P}}_{ij} = (\Box_{s_i} + w)\delta_{ij} \ , \qquad (3.47a)$$

$$\widetilde{\mathcal{N}}_{ij} = (k\partial_k R_k - \alpha_i \, \eta \, R_k)\delta_{ij} \ , \qquad (3.47b)$$

where w can be either zero or $w = -2\Lambda_k$, and η is the anomalous dimension introduced in Eq. (3.45). As we need to compute the functionals $Q_n[W_{(m)}]$, with $W_{(m)}(z)$ defined in Eq. (3.40), it is convenient to introduce the following dimensionless threshold functions [7]

$$\Phi_n^p(w) = \frac{1}{\Gamma(n)} \int_0^\infty \frac{R^{(0)}(z) - z R^{(0)\prime}(z)}{[z + R^{(0)}(z) + w]^p} z^{n-1} \, dz \qquad (3.48a)$$

$$\widetilde{\Phi}_n^p(w) = \frac{1}{\Gamma(n)} \int_0^\infty \frac{R^{(0)}(z)}{[z + R^{(0)}(z) + w]^p} z^{n-1} \, dz \qquad (3.48b)$$

with the scalar profile function $R^{(0)}(\Box_s/k^2)$ defined through the relation $R_k(\Box_s) = k^2 R^{(0)}(\Box_s/k^2)$. At this point, the right-hand-side of Eq. (3.42) can be expanded using the heat-kernel techniques. The Renormalization Group equations for the dimensionless Newton's and cosmological constants can then be written as

$$k\partial_k g_k = \beta_g(g_k, \lambda_k; D) \ , \qquad k\partial_k \lambda_k = \beta_\lambda(g_k, \lambda_k; D). \qquad (3.49)$$

In particular the beta functions can be obtained by comparing the left and right hand-sides of the Wetterich equation, and can be cast in the following form [7]

$$\beta_g = +(D - 2 + \eta)\, g \ , \tag{3.50a}$$

$$\beta_\lambda = -(2 - \eta)\, g + \frac{g}{2(4\pi)^{D/2-1}} \left\{ 2D(D+1)\, \Phi^1_{D/2}(-2\lambda) \right. \tag{3.50b}$$
$$\left. - 8D\, \Phi^1_{D/2}(0) - \eta\, D(D+1)\, \widetilde{\Phi}^1_{D/2}(-2\lambda) \right\}.$$

Here the anomalous dimension η is given by

$$\eta = \frac{g\, B_1(\lambda)}{1 - g\, B_2(\lambda)} \tag{3.51}$$

and the functions $B_1(\lambda)$ and $B_2(\lambda)$ defined as follows

$$B_1(\lambda) = +\frac{1}{3}\frac{1}{(4\pi)^{D/2-1}} \left\{ D(D+1)\, \Phi^1_{D/2-1}(-2\lambda) - 6D(D-1)\, \Phi^2_{D/2}(-2\lambda) \right.$$
$$\left. - 4D\, \Phi^1_{D/2-1}(0) - 24\, \Phi^2_{D/2}(0) \right\} \ , \tag{3.52a}$$

$$B_2(\lambda) = -\frac{1}{6}\frac{1}{(4\pi)^{D/2-1}} \left\{ D(D+1)\, \widetilde{\Phi}^1_{D/2-1}(-2\lambda) - 6D(D-1)\, \widetilde{\Phi}^2_{D/2}(-2\lambda) \right\}. \tag{3.52b}$$

The beta functions in Eq. (3.50) describe the renormalization group flow for the gravitational interaction projected onto the Einstein-Hilbert subspace, for arbitrary spacetime dimensions D. The fixed point structure can then be studied by solving the fixed point (FP) equations

$$\beta_g(g_*, \lambda_*) = 0 \ , \qquad \beta_\lambda(g_*, \lambda_*) = 0. \tag{3.53}$$

As extensively discussed in Chap. 2, the critical exponents associated with each fixed point play a central role in determining the properties of the renormalization group flow. With the aim of studying the high-energy behavior of the quantum gravitational field, it is useful to define the critical exponents θ_i as minus the eigenvalues of the stability matrix $S_{ij} \equiv \partial_{g_j}\beta_i(\mathbf{g}_*)$, and throughout this dissertation we will use this definition.

In the case of $D = 4$ spacetime dimensions, using a sharp profile function $R_k^{(0)}(q^2/k^2) = \theta(1 - q^2/k^2)$, the beta functions (3.50) assume a very simple form [10], and the Eq. (3.53) have two fixed point solutions with $g_* \geq 0$. The first one is the usual GFP, $g_* = \lambda_* = 0$ and its scaling exponents are the canonical ones. This means that, due to the negative mass dimension of the Newton's constant, the GFP is a saddle point for the gravitational renormalization group flow. The second solution of (3.53) is a non-trivial fixed point with coordinates

$$g_* = 0.403 \ , \qquad \lambda_* = 0.330. \tag{3.54}$$

This NGFP is characterized by a complex pair of critical exponents [10]

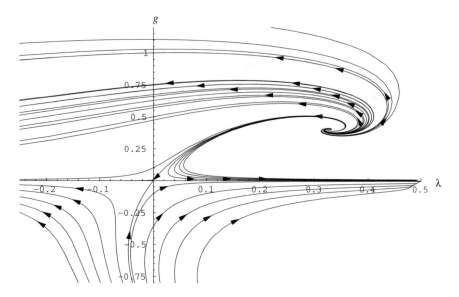

Fig. 3.1 Flow diagram for the Einstein-Hilbert truncation obtained in [10]. The arrows indicate the evolution of the renormalization group flow for decreasing values of the RG time $\tau = \log k$. In the high-energy regime the physical part of the flow ($g_k > 0$) is controlled by the NGFP, while the infrared behavior depends upon the particular RG trajectory and allows to classify the physical trajectories according to (3.56)

$$\theta_{1,2} = 1.941 \pm 3.147i \tag{3.55}$$

and since $\text{Re}(\theta_{1,2}) > 0$, the non-trivial fixed point (3.54) is attractive in the ultraviolet limit. i.e. the renormalization group flow approaches the NGFP for $k \to \infty$. In order to study the global behavior of the renormalization group flow, the Eqs. (3.49) have to be solved numerically. The RG trajectories can be obtained as the parametric curves (λ_k, g_k), with the renormalization group scale k as parameter. The resulting phase portrait [10], shown in Fig. 3.1, allows to visualize the long-term (infrared and ultraviolet) behavior of the system (3.49) for a representative set of its solutions. The interesting, physical part of the flow diagram corresponds to the half-plane $g_k > 0$. The infrared behavior ($k \to 0$) of the renormalization group flow depends on the particular trajectory. The physical RG trajectories can thus be classified according to their infrared behavior [10]

$$
\begin{array}{lll}
\text{Type Ia} & \lim_{k \to 0}(\lambda_k, g_k) = (-\infty, 0) & \Lambda_0 < 0 \\[2ex]
\text{Type IIa} & \lim_{k \to 0}(\lambda_k, g_k) = (0, 0) & \Lambda_0 = 0 \\[2ex]
\text{Type IIIa} & \lim_{k \to k_t}(\lambda_k, g_k) = (\lambda_{k_t}, g_{k_t}) & \Lambda_{k_t} > 0
\end{array}
\tag{3.56}
$$

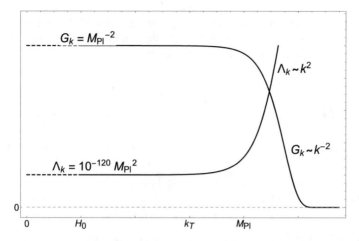

Fig. 3.2 Qualitative behavior of the running Newton's coupling and cosmological constant as functions of the RG scale k. In the high-energy regime the flow is governed by the NGFP and the couplings vary with k. By lowering the energy scale k from the ultraviolet to the infrared regime, the physical RG trajectory matches the classical regime described by General Relativity. The couplings G_k and Λ_k become constants at $k \sim M_{Pl}$ and $k \sim k_T$ respectively, k_T being the "turning point" at which the trajectory is maximally close to the GFP

Here k_t stands for the value of the RG scale k at which the anomalous dimension η becomes singular. In fact, for the Type IIIa trajectories, the renormalization group flow terminates at $k_t \neq 0$, and the infrared limit is not well defined. On the other hand, Type Ia trajectories are well defined for all k, but in the infrared limit the dimensionful cosmological constant is negative, $\Lambda_0 < 0$. At last, the only Type IIa trajectory is the "separatrix line" starting from the NGFP and terminating into the GFP for $k \to 0$. More generally a separatrix is a curve connecting different fixed points and represents the boundary separating qualitatively different behaviors in the solutions of a dynamical system. In the case at hand the separatrix (Type IIa) curve separates the Type Ia and IIIa families of solutions, which in fact differ from each other in their infrared behaviors, and correspond to different cosmologies. Notably, as the observed cosmological constant is positive, the physical trajectory describing the evolution of our universe must be a Type IIIa solution [43]. The corresponding running Newton's constant G_k and cosmological constants Λ_k are depicted in Fig. 3.2 [43]. In the Quantum Gravity regime, $k \gtrsim M_{Pl}$, the renormalization group flow is controlled by the NGFP and the running couplings scale as

$$G_k \sim k^{-2} , \qquad \Lambda_k \sim k^2. \tag{3.57}$$

Notably, one of the most important predictions of Asymptotic Safety is that the Newton's constant vanishes in the ultraviolet limit [44]. By lowering the RG scale k, the trajectory approaches the GFP and at the "turning point" $k \sim k_T$ it enters into the classical regime. Hence, in the final part of the RG evolution, the couplings G_k

and Λ_k stop their running and become constants. In particular, one can choose the RG trajectory so that the infrared (constant) values of the Newton's constant G_0 and cosmological constant Λ_0 match the observed ones.

References

1. F.J. Wegner, A. Houghton, Renormalization group equation for critical phenomena. Phys. Rev. A **8**, 401–412 (1973). https://doi.org/10.1103/PhysRevA.8.401 (cit. on p. 35)
2. K.G. Wilson, J. Kogut, The renormalization group and the ϵ expansion. Phys. Rep. **12**, 75–199 (1974). https://doi.org/10.1016/0370-1573(74)90023-4 (cit. on p. 35)
3. S. Weinberg, Critical phenomena for field theorists, in *Proceedings 14th International School of Subnuclear Physics* (Erice, 1976), p. 1. https://doi.org/10.1007/978-1-4684-0931-4_1 (cit. on p. 36)
4. C. Wetterich, Exact evolution equation for the effective potential. Phys. Lett. B **301**, 90–94 (1993). https://doi.org/10.1016/0370-2693(93)90726-X (cit. on pp. 36, 37)
5. T.R. Morris, The Exact renormalization group and approximate solutions. Int. J. Mod. Phys. A **9**, 2411–2449 (1994). https://doi.org/10.1142/S0217751X94000972. arXiv:9308265 [hep-ph] (cit. on pp. 36, 37)
6. M. Reuter, C. Wetterich, Effective average action for gauge theories and exact evolution equations. Nucl. Phys. B **417**, 181–214 (1994). https://doi.org/10.1016/0550-3213(94)90543-6 (cit. on p. 36)
7. M. Reuter, Nonperturbative evolution equation for quantum gravity. Phys. Rev. D **57**, 971–985 (1998). https://doi.org/10.1103/PhysRevD.57.971. arXiv:9605030 [hep-ph] (cit. on pp. 36, 47, 48)
8. W. Souma, Non-trivial ultraviolet fixed point in quantum gravity. Prog. Theor. Phys. **102**, 181–195 (1999). https://doi.org/10.1143/PTP.102.181. arXiv:9907027 [hep-ph] (cit. on p. 36)
9. O. Lauscher, M. Reuter, Ultraviolet fixed point and generalized flow equation of quantum gravity. Phys. Rev. D **65**(2), 025013 (2002), https://doi.org/10.1103/PhysRevD.65.025013. arXiv:0108040 [hep-th] (cit. on p. 36)
10. M. Reuter, F. Saueressig, Renormalization group flow of quantum gravity in the Einstein-Hilbert truncation. Phys. Rev. D **65**(6), 065016 (2002). https://doi.org/10.1103/PhysRevD.65.065016. arXiv:0110054 [hep-th] (cit. on pp. 36, 48–50)
11. D.F. Litim, Fixed points of quantum gravity. Phys. Rev. Lett. **92**(20), 201301 (2004). https://doi.org/10.1103/PhysRevLett.92.201301. arXiv:0312114 [hep-th] (cit. on p. 36)
12. O. Lauscher, M. Reuter, Flow equation of quantum Einstein gravity in a higher-derivative truncation. Phys. Rev. D **66**(2), 025026 (2002). https://doi.org/10.1103/PhysRevD.66.025026. arXiv:0205062 [hep-th] (cit. on p. 36)
13. A. Codello, R. Percacci, C. Rahmede, Ultraviolet properties of f(R)-gravity. Int. J. Mod. Phys. A **23**, 143–150 (2008). https://doi.org/10.1142/S0217751X08038135. arXiv:0705.1769 [hep-th] (cit. on p. 36)
14. P.F. Machado, F. Saueressig, On the renormalization group flow of f(R)-gravity. Phys. Rev. D **77**, 124045 (2008). https://doi.org/10.1103/PhysRevD.77.124045. arXiv:0712.0445 (cit. on p. 36)
15. A. Codello, R. Percacci, C. Rahmede, Investigating the ultraviolet properties of gravity with a Wilsonian renormalization group equation. Ann. Phys. **324**, 414–469 (2009). https://doi.org/10.1016/j.aop.2008.08.008. arXiv:0805.2909 [hep-th] (cit. on pp. 36, 42, 44)
16. K. Falls et al, Further evidence for asymptotic safety of quantum gravity. Phys. Rev. D **93**(10), 104022 (2016). https://doi.org/10.1103/PhysRevD.93. (cit. on p. 36)
17. M. Demmel, F. Saueressig, O. Zanusso, RG flows of quantum Einstein gravity in the linear-geometric approximation. Ann. Phys. **359**, 141–165 (2015). https://doi.org/10.1016/j.aop.2015.04.018. arXiv:1412.7207 [hep-th] (cit. on p. 36)

18. A. Codello, R. Percacci, Fixed points of higher-derivative gravity. Phys. Rev. Lett. **97**(22), 221301 (2006). https://doi.org/10.1103/PhysRevLett.97.221301. arXiv:0607128 [hep-th] (cit. on p. 36)

19. D. Benedetti, P.F. Machado, F. Saueressig, Asymptotic safety in higher- derivative gravity. Mod. Phys. Lett. A **24**, 2233–2241 (2009). https://doi.org/10.1142/S0217732309031521. arXiv:0901.2984 [hep-th] (cit. on p. 36)

20. D. Benedetti, P.F. Machado, F. Saueressig, Taming perturbative divergences in asymptotically safe gravity. Nucl. Phys. B **824** 168–191 (2010). https://doi.org/10.1016/j.nuclphysb.2009.08.023. arXiv:0902.4630 [hep-th] (cit. on p. 36)

21. F. Saueressig et al., Higher derivative gravity from the universal renormalization group machine, in *PoS EPS-HEP2011* (2011), p. 124. arXiv:1111.1743 [hep-th] (cit. on p. 36)

22. D. Benedetti, F. Caravelli, The local potential approximation in quantum gravity. J. High Energy Phys. **6**, 17 (2012). https://doi.org/10.1007/JHEP06(2012)017. arXiv:1204.3541 [hep-th] (cit. on p. 36)

23. M. Demmel, F. Saueressig, O. Zanusso, Fixed-functionals of threedimensional quantum Einstein gravity. J. High Energy Phys. **11**, 131 (2012). https://doi.org/10.1007/JHEP11(2012)131. arXiv:1208.2038 [hep-th] (cit. on p. 36)

24. J.A. Dietz, T.R. Morris. Asymptotic safety in the f(R) approximation. J. High Energy Phys. **1**, 108 (2013). https://doi.org/10.1007/JHEP01(2013)108. arXiv:1211.0955 [hep-th] (cit. on p. 36)

25. M. Demmel, F. Saueressig, O. Zanusso, Fixed functionals in asymptotically safe gravity, in *Proceedings: 13th Marcel Grossmann Meeting* (Stockholm, Sweden, 2015), pp. 2227–2229. https://doi.org/10.1142/9789814623995_0404. arXiv:1302.1312 [hep-th] (cit. on p. 36)

26. J.A. Dietz, T.R. Morris, Redundant operators in the exact renormalisation group and in the f(R) approximation to asymptotic safety. J. High Energy Phys. **7**(64), 64 (2013). https://doi.org/10.1007/JHEP07(2013)064 (cit. on p. 36)

27. D. Benedetti, F. Guarnieri, Brans-Dicke theory in the local potential approximation. New J. Phys. **16**(5), 053051 (2014). https://doi.org/10.1088/1367-2630/16/5/053051. arXiv:1311.1081 [hep-th] (cit. on p. 36)

28. M. Demmel, F. Saueressig, O. Zanusso, RG flows of quantum Einstein gravity on maximally symmetric spaces. J. High Energy Phys. **6**, 26 (2014). https://doi.org/10.1007/JHEP06(2014)026. arXiv:1401.5495 [hep-th] (cit. on p. 36)

29. R. Percacci, G.P. Vacca, Search of scaling solutions in scalar-tensor gravity. Eur. Phys. J. C **75**, 188 (2015). https://doi.org/10.1140/epjc/s10052-015-3410-0. arXiv:1501.00888 [hep-th] (cit. on p. 36)

30. J. Borchardt, B. Knorr, Global solutions of functional fixed point equations via pseudospectral methods. Phys. Rev. D **91**(10), 105011 (2015). https://doi.org/10.1103/PhysRevD.91.105011. arXiv:1502.07511 [hep-th] (cit. on p. 36)

31. M. Demmel, F. Saueressig, O. Zanusso, A proper fixed functional for four-dimensional quantum Einstein gravity. J. High Energy Phys. **8**, 113 (2015). https://doi.org/10.1007/JHEP08(2015)113. arXiv:1504.07656 [hep-th] (cit. on p. 36)

32. N. Ohta, R. Percacci, G.P. Vacca, Flow equation for f(R) gravity and some of its exact solutions. Phys. Rev. D **92**(6), 061501 (2015). https://doi.org/10.1103/PhysRevD.92.061501. arXiv:1507.00968 [hep-th] (cit. on p. 36)

33. N. Ohta, R. Percacci, G.P. Vacca, Renormalization group equation and scaling solutions for f(R) gravity in exponential parametrization. Eur. Phys. J. C **76**, 46 (2016), p. 46. https://doi.org/10.1140/epjc/s10052-016-3895-1. arXiv:1511.09393 [hep-th] (cit. on p. 36)

34. P. Labus, T.R. Morris, Z.H. Slade, Background independence in a background dependent renormalization group. Phys. Rev. D **94**(2), 024007 (2016). https://doi.org/10.1103/PhysRevD.94.024007. arXiv:1603.04772 [hep-th] (cit. on p. 36)

35. J.A. Dietz, T.R. Morris, Z.H. Slade, Fixed point structure of the conformal factor field in quantum gravity. Phys. Rev. D **94**(12), 124014 (2016). https://doi.org/10.1103/PhysRevD.94.124014. arXiv:1605.07636 [hep-th] (cit. on p. 36)

36. L.H. Ryder, *Quantum Field Theory* (Cambridge University Press, 1985). ISBN 9780521237642 (cit. on p. 37)
37. L.D. Faddeev, V.N. Popov, Feynman diagrams for the Yang-Mills field. Phys. Lett. B **25**, 29–30 (1967). https://doi.org/10.1016/0370-2693(67)90067-6 (cit. on p. 39)
38. M.E. Peskin, D.V. Schroeder. *An Introduction to Quantum Field Theory* (Addison-Wesley, Reading, USA, 1995). ISBN 9780201503975 (cit. on p. 39)
39. B.S. Dewitt, Quantum theory of gravity. II. The manifestly covariant theory. Phys. Rev. **162**, 1195–1239 (1967). https://doi.org/10.1103/PhysRev.162.1195 (cit. on p. 40)
40. L.F. Abbott, Introduction to the background field method. Acta Phys. Polon. B **13**, 33 (1982) (cit. on p. 40)
41. D. Benedetti et al., The universal RG machine. J. High Energy Phys. **6**, 79 (2011). https://doi.org/10.1007/JHEP06(2011)079. arXiv:1012.3081 [hep-th] (cit. on pp. 42, 43, 45)
42. P.B. Gilkey, The spectral geometry of a Riemannian manifold. J. Differ. Geom. **10**(4), 601–618 (1975). https://doi.org/10.4310/jdg/1214433164 (cit. on pp. 45, 46)
43. A. Bonanno, M. Reuter, Entropy signature of the running cosmological constant. J. Cosmol. Astrpart. Phys. **8**, 024 (2007). https://doi.org/10.1088/1475-7516/2007/08/024. arXiv:0706.0174 [hep-th] (cit. on p. 51)
44. A. Bonanno, M. Reuter, Renormalization group improved black hole spacetimes. Phys. Rev. D **62**(4), 043008 (2000). https://doi.org/10.1103/PhysRevD.62.043008. arXiv:0002196 [hep-th] (cit. on p. 51)

Part II
Asymptotically Safe Gravity on Foliated Spacetimes

Chapter 4
Quantum Gravity on Foliated Spacetimes

In Chap. 3 of this dissertation we introduced the metric formulation of the Asymptotic Safety scenario for Quantum Gravity. Although General Relativity requires the spacetime to be Lorentzian, the EAA is defined by means of a Euclidean path integral. In the context of Quantum Field Theory, the Lorentzian signature can be recovered by Wick-rotating all time-like quantities. However, while Quantum Field Theory is defined on a fixed Minkowski background, which furnishes a natural notion of time, the concept of time in a dynamical (and possibly fluctuating) spacetime becomes rather involved. In General Relativity the dynamical metric describing gravity is the same field defining the coordinate system. What is then the role of time in a dynamical theory of the spacetime? A way to address this question in General Relativity is the Arnowitt-Deser-Misner (ADM) formalism [1, 2]. In this construction the spacetime metric is decomposed into a lapse function N, a shift vector N_i, and a metric σ_{ij} which measures distances on the spatial slices Σ_t, defined as hypersurfaces where the time-variable t is constant. Since the ADM-formalism imprints spacetime with a foliation structure, the resulting distinguished time direction allows to compute transition amplitudes from an initial to a final slice.

A FRGE for the effective average action [3–6] tailored to the ADM-formalism has been constructed in [7, 8]. A first evaluation of the resulting renormalization group flow within the Matsubara-formalism provided strong indications that the UV fixed point underlying the Asymptotic Safety scenario is robust under a change from Euclidean to Lorentzian signature [7, 8].

This chapter is based on the following publications:

• J. Biemans, A. Platania, F. Saueressig-*Quantum gravity on foliated spacetimes-Asymptotically safe and sound*-Phys. Rev. D 95 (2017) 086013 [arXiv:1609.04813]

• J. Biemans, A. Platania, F. Saueressig-*Renormalization group fixed points of foliated gravity-matter systems*-JHEP 05 (2017) 093 [arXiv:1702.06539]

• A. Platania, F. Saueressig-*Functional Renormalization Group flows on Friedman-Lemaitre-Robertson-Walker backgrounds* Found. Phys. (2018). https://doi.org/10.1007/s10701-018-0181-0.

In this chapter we employ the ADM-formalism for evaluating the gravitational renormalization group flow on a cosmological Friedmann-Robertson-Walker (FRW) background. In particular we will show that the renormalization group flow resulting from projecting the FRGE onto the Einstein-Hilbert action exhibits a NGFP suitable for Asymptotic Safety. Furthermore, given that asymptotic safety is a rather general concept whose applicability is not limited to the gravitational interaction, it is also interesting to study the renormalization group flow of gravity minimally coupled to matter fields. In the context of the metric formulation of Quantum Gravity, it has been shown that the Asymptotic Safety mechanism may also play a key role in the high-energy completion of a large class of gravity-matter models [9–22]. Thus the final part of the present chapter is devoted to the study of matter effects in the ADM setting. In particular we will discuss the fixed point structure arising from foliated gravity-matter systems containing an arbitrary number of minimally coupled scalars, N_S, vector fields N_V, and Dirac fermions N_D.

The inclusion of the matter fields leads to a two-parameter deformation of the beta functions controlling the flow of G_k and Λ_k. Analyzing the beta functions of the gravity-matter systems utilizing these deformation parameters allows classifying their fixed point structure. The fixed points for a specific gravity-matter model can then be determined by evaluating the map relating its field content to the deformation parameters. In particular, we will see that the matter content of the Standard Model of particle physics as well as many of its phenomenologically motivated extensions are located in areas which give rise to a single UV fixed point with real critical exponents. These findings provide a first indication that the asymptotic safety mechanism encountered in the case of pure gravity may carry over to the case of foliated gravity-matter systems with a realistic matter content.

4.1 Spacetime Foliation and Functional Renormalization

The FRGE on foliated spacetimes has been constructed in [7, 8] and we review the formalism in the following sections.

4.1.1 Arnowitt-Deser-Misner Decomposition of Spacetime

Let us consider a D-dimensional Euclidean manifold $\mathcal{M}$ with metric $g_{\mu\nu}$, carrying coordinates x^α. In order to set up a preferred "time"-direction, we define a time function $\tau(x)$ which assigns a specific time τ to each spacetime point x. This can be used to decompose $\mathcal{M}$ into a stack of spatial slices $\Sigma_{\tau_i} \equiv \{x : \tau(x) = \tau_i\}$ encompassing all points x with the same value of the "time-coordinate" τ_i. The gradient of the time function $\partial_\mu \tau$ can be used to define a vector n^μ normal to the spatial slices, $n_\mu \equiv N \partial_\mu \tau$. The lapse function $N(\tau, y^i)$ ensures the normalization $g_{\mu\nu} n^\mu n^\nu = 1$. Furthermore, the gradient can be used to introduce a vector field t^μ satisfying $t^\mu \partial_\mu \tau = 1$. Denoting the coordinates on Σ_τ by y^i, $i = 1, \ldots, d$, the tangent space on a point in $\mathcal{M}$ can then be decomposed into the space tangent to Σ_τ and its complement. The corresponding basis vectors can be constructed from the Jacobians

$$t^\mu = \left.\frac{\partial x^\mu}{\partial \tau}\right|_{y^i}, \qquad e_i{}^\mu = \left.\frac{\partial x^\mu}{\partial y^i}\right|_\tau. \tag{4.1}$$

The normal vector then satisfies $g_{\mu\nu} n^\mu e_i{}^\nu = 0$.

The spatial coordinate systems on neighboring spatial slices can be connected by constructing the integral curves of t^μ and requiring that y^i is constant along these curves. A priori t^μ is neither tangent nor orthogonal to the spatial slices. Using the Jacobians (4.1) it can be decomposed into its components normal and tangent to Σ_τ

$$t^\mu = N n^\mu + N^i e_i{}^\mu, \tag{4.2}$$

where $N^i(\tau, y^i)$ is called shift vector. Analogously, the coordinate one-forms transform according to

$$dx^\mu = t^\mu d\tau + e_i{}^\mu dy^i = N n^\mu d\tau + e_i{}^\mu (dy^i + N^i d\tau). \tag{4.3}$$

Defining the metric on the spatial slice $\sigma_{ij} = e_i{}^\mu e_j{}^\nu g_{\mu\nu}$, the line-element $ds^2 = g_{\mu\nu} dx^\mu dx^\nu$ written in terms of the ADM fields takes the form

$$ds^2 = g_{\alpha\beta} dx^\alpha dx^\beta = N^2 d\tau^2 + \sigma_{ij} (dy^i + N^i d\tau)(dy^j + N^j d\tau). \tag{4.4}$$

Note that in this case the lapse function N, the shift vector N^i and the induced metric on the spatial slices σ_{ij} depend on the spacetime coordinates (τ, y^i).[1] In terms of metric components, the decomposition (4.4) implies

$$g_{\alpha\beta} = \begin{pmatrix} N^2 + N_i N^i & N_j \\ N_i & \sigma_{ij} \end{pmatrix}, \qquad g^{\alpha\beta} = \begin{pmatrix} \frac{1}{N^2} & -\frac{N^j}{N^2} \\ -\frac{N^i}{N^2} & \sigma^{ij} + \frac{N^i N^j}{N^2} \end{pmatrix} \tag{4.5}$$

where spatial indices i, j are raised and lowered with the metric on the spatial slices. An infinitesimal coordinate transformation $v^\alpha(\tau, y)$ acting on the metric can be expressed in terms of the Lie derivative $\mathcal{L}_v$

$$\delta g_{\alpha\beta} = \mathcal{L}_v g_{\alpha\beta}. \tag{4.6}$$

Decomposing

$$v^\alpha = \big(f(\tau, y), \zeta^i(\tau, y)\big) \tag{4.7}$$

---

[1]This situation differs from projectable Hořava-Lifshitz gravity where N is restricted to be a function of (Euclidean) time τ only.

into its temporal and spatial parts, the transformation (4.6) determines the transformation properties of the component fields under Diff($\mathcal{M}$)

$$
\begin{aligned}
\delta N &= \partial_\tau (fN) + \zeta^k \partial_k N - N N^i \partial_i f \,, \\
\delta N_i &= \partial_\tau (N_i f) + \zeta^k \partial_k N_i + N_k \partial_i \zeta^k + \sigma_{ki} \partial_\tau \zeta^k + N_k N^k \partial_i f + N^2 \partial_i f \,, \quad (4.8) \\
\delta \sigma_{ij} &= f \partial_\tau \sigma_{ij} + \zeta^k \partial_k \sigma_{ij} + \sigma_{jk} \partial_i \zeta^k + \sigma_{ik} \partial_j \zeta^k + N_j \partial_i f + N_i \partial_j f \,.
\end{aligned}
$$

For completeness, we note

$$
\delta N^i = \partial_\tau (N^i f) + \zeta^j \partial_j N^i - N^j \partial_j \zeta^i + \partial_\tau \zeta^i - N^i N^j \partial_j f + N^2 \sigma^{ij} \partial_j f \,. \quad (4.9)
$$

Denoting expressions in Euclidean and Lorentzian signature by subscripts E and L, the Wick rotation is implemented by

$$
\tau_E \to -i \tau_L \,, \qquad N^i_E \to i N^i_L \,. \quad (4.10)
$$

In the case where τ is associated with a Killing vector, this transformation allows to go from Euclidean to Lorentzian metrics.

The (Euclidean) Einstein-Hilbert action written in ADM fields reads

$$
S^{\mathrm{EH}} = \frac{1}{16\pi G} \int d\tau d^d y \, N \sqrt{\sigma} \left[K_{ij} \mathcal{G}^{ij,kl} K_{kl} - {}^{(d)}R + 2\Lambda \right] \,. \quad (4.11)
$$

Here ${}^{(d)}R$ denotes the intrinsic curvature on the d-dimensional spatial slice,

$$
K_{ij} \equiv \frac{1}{2N} \left(\partial_\tau \sigma_{ij} - D_i N_j - D_j N_i \right) \,, \qquad K \equiv \sigma^{ij} K_{ij} \quad (4.12)
$$

are the extrinsic curvature and its trace, and D_i denotes the covariant derivative constructed from σ_{ij}. The kinetic term is determined by the Wheeler-de Witt metric

$$
\mathcal{G}^{ij,kl} \equiv \sigma^{ik} \sigma^{jl} - \lambda \sigma^{ij} \sigma^{kl} \,. \quad (4.13)
$$

The parameter $\lambda = 1$ is fixed by requiring invariance of the action with respect to Diff($\mathcal{M}$) and, in what follows, we adhere to this value.

4.1.2 Renormalization Group Equations on Foliated Spacetimes

In Chap. 3 we reviewed the FRG in the metric approach to Quantum Gravity [6]. The study of the gravitational renormalization group flow on foliated spacetimes requires a formulation of the FRGE

$$k\partial_k \Gamma_k[h; \bar{g}] = \frac{1}{2} \text{STr} \left[\left(\Gamma_k^{(2)} + \mathcal{R}_k \right)^{-1} k\partial_k \mathcal{R}_k \right] \tag{4.14}$$

where the (Euclidean) spacetime metric $g_{\mu\nu}$ is decomposed according to (4.4), and the gravitational degrees of freedom are encoded in the ADM-fields $\{N, N_i, \sigma_{ij}\}$ [7, 8]. The effective average action Γ_k is then obtained in the usual way. In particular, the construction of Γ_k makes manifest use of the BFM, introduced in Sect. 3.2.2. Following [8] we use a linear split of the ADM fields into background fields (marked with an bar) and fluctuations (indicated by a hat)[2]

$$N = \bar{N} + \hat{N}, \qquad N_i = \bar{N}_i + \hat{N}_i, \qquad \sigma_{ij} = \bar{\sigma}_{ij} + \hat{\sigma}_{ij}. \tag{4.15}$$

Owed to the non-linearity of the ADM decomposition, the transformation of the ADM fields under the full diffeomorphism group is non-linear. In combination with the linear split (4.15) this entails that $\Delta_k S$, which, by construction, is quadratic in the fluctuation fields, preserves a subgroup of the full diffeomorphism group as a background symmetry only. Inspecting Eqs. (4.8) and (4.9) one sees that restricting the symmetry group to foliation preserving diffeomorphisms where, by definition, $f(\tau, y) = f(\tau)$ is independent of the spatial coordinates, eliminates the quadratic terms in the transformation laws. This indicates that the regulator appearing in (4.14), as well as the gauge-fixing and ghost contributions, only respects foliation preserving diffeomorphisms as a background symmetry. The resulting FRGE retains foliation-preserving background diffeomorphisms as a background symmetry only (see [8] for a detailed discussion).

Following the notation introduced in Chap. 3, we will denote the sets of physical fields, background fields and fluctuations by χ, $\bar{\chi}$, and $\hat{\chi}$, respectively. Accordingly, the generating functional has the form (3.20), namely

$$Z_k[J; \bar{\chi}] \equiv \int \mathcal{D}[\hat{\chi}] \exp\left[-S^{\text{grav}} - S^{\text{matter}} - S^{\text{gf}} - S^{\text{ghost}} - \Delta_k S + S^{\text{source}}\right], \tag{4.16}$$

where $S^{\text{grav}}[N, N_i, \sigma_{ij}]$ is a generic diffeomorphism invariant action, supplemented by a suitable gauge-fixing term S^{gf}, a corresponding ghost action S^{ghost}, and source terms $S^{\text{source}} \equiv J \cdot \hat{\chi}$ for the fluctuation fields. In addition to the gravitational action $S^{\text{grav}}[N, N_i, \sigma_{ij}]$, we also consider N_S scalar fields, N_V abelian gauge fields and N_D Dirac fields minimally coupled to gravity through the action

$$S^{\text{matter}} = S^{\text{scalar}} + S^{\text{vector}} + S^{\text{fermion}}, \tag{4.17}$$

where

---

[2] Strictly speaking, the fields appearing in the EAA are the vacuum expectation values of the classical fields introduced in the previous subsection. In order to keep our notation light, we use the same notation for both fields, expecting that the precise meaning is clear from the context.

$$
S^{\text{scalar}} = \frac{1}{2} \sum_{i=1}^{N_S} \int d\tau d^d x N \sqrt{\sigma} \left[\phi^i \, \Delta_0 \, \phi^i \right] ,
$$

$$
S^{\text{vector}} = \frac{1}{4} \sum_{i=1}^{N_V} \int d\tau d^d x N \sqrt{\sigma} \left[g^{\mu\nu} g^{\alpha\beta} F^i_{\mu\alpha} F^i_{\nu\beta} \right] + \frac{1}{2\xi} \sum_{i=1}^{N_V} \int d\tau d^d x \bar{N} \sqrt{\sigma} \left[\bar{g}^{\mu\nu} \bar{D}_\mu A^i_\nu \right]^2
$$

$$
+ \sum_{i=1}^{N_V} \int d\tau d^d x \bar{N} \sqrt{\bar{\sigma}} \left[\bar{C}^i \, \Delta_0 \, C^i \right] ,
$$

$$
S^{\text{fermion}} = i \sum_{i=1}^{N_D} \int d\tau d^d x N \sqrt{\sigma} \left[\bar{\psi}^i \, \slashed{\nabla} \, \psi^i \right] . \tag{4.18}
$$

The summation index i runs over the matter species. Moreover, we adopt Feynman gauge for the vector fields, setting $\xi = 1$. In the context of Asymptotic Safety, matter sectors of this type have been discussed in covariant approach in [9, 10] with extensions considered recently in [15, 17, 19, 23]. In particular, our treatment of the Dirac fermions follows [15, 24]. All matter actions are finally converted to the ADM framework by using the projector (4.1).

At this stage it is instructive to contrast the background field formalism set up in terms of ADM-variables with the covariant field decompositions discussed in [25, 26]. We start by considering a linear split of the spacetime metric $g_{\mu\nu}$ into background $\bar{g}_{\mu\nu}$ and fluctuations $h_{\mu\nu}$

$$
g_{\mu\nu} = \bar{g}_{\mu\nu} + h_{\mu\nu} . \tag{4.19}
$$

Applying the ADM-decomposition (4.5) to $\bar{g}_{\mu\nu}$ expresses $\bar{g}_{\mu\nu}$ in terms of the background ADM-fields $(\bar{N}, \bar{N}_i, \bar{\sigma}_{ij})$. Performing the same decomposition for $g_{\mu\nu}$ and subsequently substituting the linear decomposition of the ADM-fields (4.15) then provides a relation between the fluctuations $h_{\mu\nu}$ and the fields appearing in the ADM-formulation

$$
\begin{aligned}
h_{00} &= 2\bar{N}\hat{N} + \hat{N}^2 + \sigma^{ij}(\bar{N}_i + \hat{N}_i)(\bar{N}_j + \hat{N}_j) - \bar{\sigma}^{ij}\bar{N}_i\bar{N}_j , \\
h_{0i} &= \hat{N}_i , \\
h_{ij} &= \hat{\sigma}_{ij} .
\end{aligned} \tag{4.20}
$$

The relations containing spatial indices are linear while the expression for h_{00} involves both the background and fluctuating ADM fields to arbitrary high powers. This is reminiscent of the exponential parameterization of the metric fluctuation which also involves $h_{\mu\nu}$ to arbitrary high powers. The map (4.20) then establishes that the ADM-decomposition gives rise to a natural parameterization of the metric fluctuations.

4.2 Renormalization Group Flow on a FRW Background

In this section we use the flow equation to study the gravitational RG flow projected on the ADM-decomposed Einstein-Hilbert action evaluated on a flat FRW background [27]. In this case the flow of the cosmological constant and Newton's constant is encoded in the volume factor and extrinsic curvature terms constructed from the background. This bypasses one of the main limitations of the Matsubara-type computations [7, 8] where time-direction was taken compact. We will start our discussion by considering the case of pure gravity, while the case of gravity minimally coupled to non-interacting matter fields will be studied in Sect. 4.3.2.

4.2.1 The Einstein-Hilbert Ansatz

Finding exact solutions of the FRGE (4.14) is rather difficult. As discussed in the previous chapters, a standard way of constructing approximate solutions is to restrict the interaction monomials in Γ_k to a specific subset and subsequently project the renormalization group flow onto the subspace spanned by the ansatz.

In the present discussion we focus on the study of the renormalization group flow projected onto the Einstein-Hilbert action in the ADM-formalism, Eq. (4.11). The gravitational part of the EAA thus reads

$$\Gamma_k^{\text{grav}} \simeq \frac{1}{16\pi G_k} \int d\tau d^d y \, N \sqrt{\sigma} \left[K_{ij} K^{ij} - K^2 - {}^{(d)}R + 2\Lambda_k \right]. \tag{4.21}$$

This ansatz contains two scale-dependent couplings, the Newton's constant G_k and the cosmological constant Λ_k. Since the background field is arbitrary, it can be chosen in a way that facilitate the computation. For the ansatz (4.21) it then suffices to evaluate the flow on a flat (Euclidean) Friedmann-Robertson-Walker (FRW) background

$$\bar{g}_{\mu\nu} = \text{diag}\left[1, \, a(\tau)^2 \, \delta_{ij} \right] \quad \Longleftrightarrow \quad \bar{N} = 1, \quad \bar{N}_i = 0, \quad \bar{\sigma}_{ij} = a(\tau)^2 \, \delta_{ij}, \tag{4.22}$$

where $a(\tau)$ is a positive, time-dependent scale factor. In this background the projectors (4.1) take a particularly simple form

$$t^\mu = \left(1, \vec{0} \right), \qquad e_i{}^\mu = \left(\vec{0}, \, \delta_i^j \right), \tag{4.23}$$

implying that t^μ is always normal to the spatial hypersurface Σ_τ. The extrinsic and intrinsic curvature tensors of this background satisfy

$$\bar{K}_{ij} = \frac{1}{d} \bar{K} \, \bar{\sigma}_{ij}, \qquad {}^{(d)}\bar{R} = 0 \tag{4.24}$$

where $\bar{K} \equiv \bar{\sigma}^{ij} \bar{K}_{ij}$. Moreover, the Christoffel connection on the spatial slices vanishes such that $\bar{D}_i = \partial_i$. Evaluating (4.21) on this background using (4.24) yields

$$\Gamma_k^{\mathrm{grav}}\big|_{\hat{\chi}=0} = \frac{1}{16\pi G_k} \int d\tau d^d y \sqrt{\bar{\sigma}} \left[-\tfrac{d-1}{d} \bar{K}^2 + 2\Lambda_k \right], \qquad (4.25)$$

where $\hat{\chi}$ denotes the set of all fluctuation fields. Thus the choice (4.22) is sufficiently general to distinguish the two interaction monomials encoding the flow of G_k and Λ_k. Note that we have not assumed that the background is compact. In particular the "time-coordinate" τ may be taken as non-compact.

4.2.2 Operator Traces on FRW Backgrounds

In order to evaluate the operator traces appearing in the flow equation we need to resort to heat-kernel techniques with respect to the background spacetime (4.22). For this purpose, we observe that (4.23) entails that there is a canonical "lifting" of vectors tangent to the spatial slice to D-dimensional vectors

$$v^i(\tau, y) \quad \mapsto \quad v^\mu(\tau, y) \equiv (0, \, v^i(\tau, y))^{\mathrm{T}}. \qquad (4.26)$$

The D-dimensional Laplacian $\Box_s \equiv -\bar{g}^{\mu\nu} \bar{D}_\mu \bar{D}_\nu$ $(s = 0, 1, 2)$ naturally acts on these D-vectors. In order to rewrite the variations in terms of D-covariant quantities, we exploit that $\Box_s$ can be expressed in terms of the flat space Laplacian $\Box \equiv -\partial_\tau^2 - \bar{\sigma}^{ij} \partial_i \partial_j$ and the extrinsic curvature. For the Laplacian acting on D-dimensional fields with zero, one, and two indices one has

$$\Box_0 \phi = \left(\Box - \bar{K} \partial_\tau \right) \phi,$$
$$\Box_1 \phi_\mu = \left(\Box - \tfrac{d-2}{d} \bar{K} \partial_\tau + \tfrac{1}{d}(\partial_\tau \bar{K}) + \tfrac{1}{d}\bar{K}^2 \right) \phi_\mu, \qquad (4.27)$$
$$\Box_2 \phi_{\mu\nu} = \left(\Box - \tfrac{d-4}{d} \bar{K} \partial_\tau + \tfrac{2}{d}(\partial_\tau \bar{K}) + \tfrac{2(d-1)}{d^2}\bar{K}^2 \right) \phi_{\mu\nu}.$$

When evaluating the traces by covariant heat-kernel methods we then use the embedding map (4.26) together with the completion (4.27) to express the operator $\Box$ in terms of $\Box_{s_i}$. In what follows we use $\Box_i \equiv \Box_{s_i}$ to shorten notation.

Table 4.1 Heat-kernel coefficients for the component fields appearing in the decompositions (4.34) and (4.35). Here S, V, T, TV, and TTT are scalars, vectors, symmetric two-tensors, transverse vectors, and transverse-traceless symmetric matrices, respectively

	S	V	T	TV	TTT
a_0	1	d	$\frac{1}{2}d(d+1)$	$d-1$	$\frac{1}{2}(d+1)(d-2)$
a_2	$\frac{d-1}{6d}$	$\frac{d-1}{6}$	$\frac{(d-1)(d+1)}{12}$	$\frac{d^3-2d^2+d+6}{6d^2}$	$\frac{d^4-2d^3-d^2+14d+36}{12d^2}$

The operator traces appearing in (4.14) are conveniently evaluated using standard heat-kernel formulas for the D-dimensional Laplacians (4.27). Following the notation introduced in Sect. 3.2.3, the early-time expansion (3.36) yields

$$\text{Tr}\, e^{-s\left(\Box_i+\hat{E}\right)} \simeq \frac{1}{(4\pi s)^{D/2}} \int \sqrt{g}\, \left[\text{Tr}\, \hat{1}_i + s\left(\tfrac{1}{6}\,{}^{(D)}R\,\text{Tr}\, \hat{1}_i - \text{Tr}\,\hat{E}\right) + \dots\right] d^D x \tag{4.28}$$

where the dots indicate terms built from four and more covariant derivatives, which do not contribute to the present computation. For the FRW background, the background curvature ${}^{(D)}\bar{R}$ can readily be replaced by the extrinsic curvature

$$\int d^D x \sqrt{g}\, {}^{(D)}\bar{R} = \int d\tau d^d y \sqrt{\bar{\sigma}}\left[\tfrac{d-1}{d}\bar{K}^2\right] . \tag{4.29}$$

Combining the diagonal form of the projectors (4.23) with the D-dimensional heat-kernel expansion (4.28) allows to write operator traces for the component fields χ_i. On the flat FRW background these have the structure

$$\text{Tr}\, e^{-s\Box_i} = \frac{1}{(4\pi s)^{D/2}} \int \sqrt{\bar{\sigma}}\left[a_0 + a_2\, s\, \bar{K}^2 + \dots\right] d\tau d^d y . \tag{4.30}$$

The coefficient a_n depend on the index structure of the fluctuation fields and are listed in Table 4.1. The result (4.30), together with Table 4.1, are the key ingredient for evaluating the operator traces of the flow Eq. (4.14) on a flat FRW background.

4.2.3 Hessians, Gauge-Fixing and Ghost Action

Constructing the right-hand-side of the flow equation requires the Hessian $\Gamma_k^{(2)}$. Starting with the contributions originating from the gravitational action Γ_k^{grav}, it is convenient to introduce the building blocks

$$\begin{aligned} I_1 &\equiv \int d\tau d^d y\, N \sqrt{\sigma}\, K_{ij}K^{ij}, & I_2 &\equiv \int d\tau d^d y\, N \sqrt{\sigma}\, K^2, \\ I_3 &\equiv \int d\tau d^d y\, N \sqrt{\sigma}\, {}^{(d)}R, & I_4 &\equiv \int d\tau d^d y\, N \sqrt{\sigma}, \end{aligned} \tag{4.31}$$

such that

$$\Gamma_k^{\text{grav}} = \frac{1}{16\pi G_k} \left(I_1 - I_2 - I_3 + 2\Lambda_k\, I_4 \right) . \tag{4.32}$$

Expanding this expression around the background (4.22), the terms quadratic in the fluctuation fields then take the form

$$\delta^2 \Gamma_k^{\text{grav}} = \frac{1}{16\pi G_k} \left(\delta^2 I_1 - \delta^2 I_2 - \delta^2 I_3 + 2\Lambda_k\, \delta^2 I_4 \right) , \tag{4.33}$$

with the explicit expressions for $\delta^2 I_i$ given in (A.5).

At this stage it convenient to express the fluctuation fields in terms of the component fields used in cosmic perturbation theory (see, e.g., [28] for a pedagogical introduction). Defining $\Delta \equiv -\bar{\sigma}^{ij}\partial_i\partial_j$, the shift vector is decomposed into its transverse and longitudinal parts according to

$$\hat{N}_i = u_i + \partial_i \frac{1}{\sqrt{\Delta}} B, \qquad \partial^i u_i = 0. \tag{4.34}$$

The metric fluctuations are written as

$$\hat{\sigma}_{ij} = h_{ij} - \left(\bar{\sigma}_{ij} + \partial_i\partial_j \frac{1}{\Delta}\right)\psi + \partial_i\partial_j \frac{1}{\Delta} E + \partial_i \frac{1}{\sqrt{\Delta}} v_j + \partial_j \frac{1}{\sqrt{\Delta}} v_i, \qquad \hat{\sigma} \equiv \bar{\sigma}^{ij}\hat{\sigma}_{ij}, \tag{4.35}$$

with the component fields subject to the differential constraints

$$\partial^i h_{ij} = 0, \qquad \bar{\sigma}^{ij} h_{ij} = 0, \qquad \partial^i v_i = 0. \tag{4.36}$$

The partial derivatives and Δ can be commuted freely, since the background metric is independent of the spatial coordinates. The normalization of the component fields has been chosen such that the change of integration variables does not give rise to non-trivial Jacobians. This can be seen from noting

$$\int_x \hat{N}_i\,\hat{N}^i = \int_x (u_i\, u^i + B^2) ,$$
$$\int_x \hat{\sigma}_{ij}\,\hat{\sigma}^{ij} = \int_x (h_{ij} h^{ij} + (d-1)\,\psi^2 + E^2 + 2\, v_i\, v^i) , \tag{4.37}$$

implying that a Gaussian integral over the ADM fluctuations leads to a Gaussian integral in the component fields which does not give rise to operator-valued determinants. The result obtained from substituting the decompositions (4.34) and (4.35) into the variations $\delta^2 I_i$ is given in Eqs. (A.10), (A.11), and (A.12). On this basis it is then rather straightforward to write down the explicit form of (4.33) in terms of the component fields.

At this stage it is important to investigate the matrix elements of $\delta^2\Gamma_k^{\text{grav}}$ on flat Euclidean space, obtained by setting $\bar{K} = 0$. The result is summarized in the second column of Table 4.2. On this basis, one can make the crucial observation that the component fields do not possess a relativistic dispersion relation. One may then

Table 4.2 Matrix elements appearing in $\delta^2\Gamma_k$ when expanding Γ_k around flat Euclidean space. The column "index" identifies the corresponding matrix element in field space, $\Delta \equiv -\bar{\sigma}^{ij}\partial_i\partial_j$ is the Laplacian on the spatial slice, and $\Box \equiv -\partial_\tau^2 - \bar{\sigma}^{ij}\partial_i\partial_j$. For each "off-diagonal" entry there is a second contribution involving the adjoint of the differential operator and the order of the fields reversed

Index	Matrix element $32\pi G_k\,\delta^2\Gamma_k^{\text{grav}}$	Matrix element $32\pi G_k\left(\delta^2\Gamma_k^{\text{grav}} + \Gamma_k^{\text{gf}}\right)$
$h\,h$	$\Box - 2\Lambda_k$	$\Box - 2\Lambda_k$
$v\,v$	$2\left[-\partial_\tau^2 - 2\Lambda_k\right]$	$\Box - 2\Lambda_k$
$E\,E$	$-\Lambda_k$	$\frac{1}{2}(\Box - 2\Lambda_k)$
$\psi\,\psi$	$-(d-1)(d-2)\left[\Box - \frac{d-3}{d-2}\,\Lambda_k\right]$	$-\frac{(d-1)(d-3)}{2}\left[\Box - 2\,\Lambda_k\right]$
$\psi\,E$	$-(d-1)\left[-\partial_\tau^2 - 2\Lambda_k\right]$	$-(d-1)\left[\Box - 2\Lambda_k\right]$
$u\,u$	$2\,\Delta$	$2\,\Box$
$u\,v$	$-2\,\partial_\tau\sqrt{\Delta}$	0
$B\,\psi$	$2\,(d-1)\sqrt{\Delta}\,\partial_\tau$	0
$\hat{N}\,\psi$	$2\,(d-1)\left[\Delta - \Lambda_k\right]$	$(d-1)\left[\Box - 2\Lambda_k\right]$
$\hat{N}\,E$	$-2\,\Lambda_k$	$\Box - 2\Lambda_k$
$\hat{N}\,\hat{N}$	0	$2\,\Box$

attempt to add a suitable gauge-fixing term Γ_k^{gf} ameliorating the structure of the kinetic part of the Hessian. Inspired by the decomposition (4.7) the gauge-fixing of the symmetries (4.8) may be implemented via two functions F and F_i

$$\Gamma_k^{\text{gf}} = \frac{1}{32\pi G_k}\int d\tau d^d y\,\sqrt{\bar{\sigma}}\,\left[F_i\,\bar{\sigma}^{ij}F_j + F^2\right], \qquad (4.38)$$

where F and F_i are linear in the fluctuation fields. The integrand entering Γ_k^{gf} may also be written in terms of a D-dimensional vector $F_\mu \equiv (F, F_i)$ and the background metric (4.22), exploiting that $F_\mu\,\bar{g}^{\mu\nu}\,F_\nu = F^2 + F_i\,\bar{\sigma}^{ij}\,F_j$. The most general form of F and F_i which is linear in the fluctuation fields $(\hat{N}, \hat{N}_i, \hat{\sigma}_{ij})$ and involves at most one derivative with respect to the spatial or time coordinate is given by

$$\begin{aligned}
F &= c_1\,\partial_\tau\,\hat{N} + c_2\,\partial^i\,\hat{N}_i + c_3\,\partial_\tau\,\hat{\sigma} + d\,c_8\,\bar{K}^{ij}\,\hat{\sigma}_{ij} + c_9\,\bar{K}\hat{N}, \\
F_i &= c_4\,\partial_\tau\,\hat{N}_i + c_5\,\partial_i\,\hat{N} + c_6\,\partial_i\,\hat{\sigma} + c_7\,\partial^j\,\hat{\sigma}_{ji} + d\,c_{10}\,\bar{K}_{ij}\hat{N}^j.
\end{aligned} \qquad (4.39)$$

The c_i are real coefficients which may depend on d and the factors d are introduced for later convenience. Following the calculation in Appendix A.1.2, the gauge-fixing (4.38) can be written in terms of the component fields (4.34) and (4.35), and the result is summarized in Eqs. (A.14) and (A.15). Combining $\delta^2\Gamma_k^{\text{grav}}$ with the gauge-fixing contribution one finally arrives at (A.16). The coefficients c_i are then fixed by requiring, firstly, that all component fields come with a relativistic dispersion relation and, secondly, that the resulting gauge-fixed Hessian does not contain square-roots of the spatial Laplacian, $\sqrt{\Delta}$. The resulting coefficients are given by

$$c_1 = \epsilon_1 \,, \ c_2 = \epsilon_1 \,, \quad c_3 = -\tfrac{1}{2}\epsilon_1 \,, \ c_8 = 0 \,, \ c_9 = \tfrac{2(d-1)}{d}\epsilon_1 \,,$$
$$c_4 = \epsilon_2 \ \ c_5 = -\epsilon_2 \,, \ c_6 = -\tfrac{1}{2}\epsilon_2 \,, \ c_7 = \epsilon_2 \,, \ c_{10} = \tfrac{d-2}{d}\epsilon_2 \,. \tag{4.40}$$

Here $\epsilon_1 = \pm 1$ and $\epsilon_2 = \pm 1$. Since Γ_k^{gf} is quadratic in F and F_i it depends on ϵ_i^2 only and the choice of sign does not change Γ_k^{gf}. Therefore, the aforementioned conditions fix the gauge *uniquely* [29], up to a physically irrelevant discrete symmetry.

Notably, the gauge fixing (4.38) bears a close relation with the de Witt (dW) background covariant gauge, which reads

$$\Gamma_k^{\mathrm{gf}} = \frac{1}{32\pi G_k} \int d^D x \sqrt{\bar{g}} \left[F_\mu^{\mathrm{dW}} \, \bar{g}^{\mu\nu} \, F_\nu^{\mathrm{dW}} \right] , \tag{4.41}$$

where the gauge-fixing function F_μ^{dW} is defined as follows

$$F_\mu^{\mathrm{dW}} \equiv \bar{D}^\nu h_{\nu\mu} - \tfrac{1}{2} \bar{D}_\mu \left(\bar{g}^{\alpha\beta} h_{\alpha\beta} \right) . \tag{4.42}$$

This becomes apparent when F_μ^{dW} is decomposed into its time and spatial parts, $F_\mu^{\mathrm{dW}} = (F^{\mathrm{dW}}, F_i^{\mathrm{dW}})$. Substituting the relations (4.20) (truncated at linear order in the fluctuation fields) and choosing a flat space background gives

$$F^{\mathrm{dW}} = \partial_\tau \hat{N} + \partial^i \hat{N}_i - \tfrac{1}{2}\partial_\tau \hat{\sigma} + O(\hat{\chi}^2) \,,$$
$$F_i^{\mathrm{dW}} = \partial_\tau \hat{N}_i - \partial_i \hat{N} - \tfrac{1}{2}\partial_i \hat{\sigma} + \partial^j \hat{\sigma}_{ji} + O(\hat{\chi}^2) \,. \tag{4.43}$$

This result coincides with (4.39). Thus it is clear that the gauge fixing (4.38) can be completed such that it preserves full background diffeomorphism symmetry by systematically starting from (4.41) and substituting the full map (4.20). Since the resulting extra terms do not contribute to the present computation, they will not be considered any further.

Combining (4.33) with the gauge choice (4.38) and (4.40) finally results in the gauge-fixed Hessian

$$32\pi G_k \left(\tfrac{1}{2}\delta^2 \Gamma_k^{\mathrm{grav}} + \Gamma_k^{\mathrm{gf}} \right) =$$
$$\int_x \Bigg\{ + \tfrac{1}{2} h^{ij} \left[\Box_2 - 2\Lambda_k - \tfrac{2(d-1)}{d}\dot{\bar{K}} - \tfrac{d^2-d+2}{d^2}\bar{K}^2 \right] h_{ij}$$
$$+ u^i \left[\Box_1 - \tfrac{d-1}{d}\dot{\bar{K}} - \tfrac{1}{d}\bar{K}^2 \right] u_i + v^i \left[\Box_1 - 2\Lambda_k - \dot{\bar{K}} - \tfrac{5d-7}{d^2}\bar{K}^2 \right] v_i$$
$$+ B \left[\Box_0 - \tfrac{d-1}{d}\dot{\bar{K}} - \tfrac{d-1}{d^2}\bar{K}^2 \right] B + \hat{N} \left[\Box_0 - \tfrac{2(d-1)}{d}\dot{\bar{K}} - \tfrac{4(d-1)}{d^2}\bar{K}^2 \right] \hat{N}$$
$$+ \hat{N} \left[\Box_0 - 2\Lambda_k - \tfrac{5d^2-12d+16}{4d^2}\bar{K}^2 \right] ((d-1)\psi + E)$$
$$+ \tfrac{1}{4} E \left[\Box_0 - 2\Lambda_k - \tfrac{2(d-1)}{d}\dot{\bar{K}} - \tfrac{d-1}{d}\bar{K}^2 \right] E$$
$$- \tfrac{(d-1)(d-3)}{4} \psi \left[\Box_0 - 2\Lambda_k - \tfrac{2(d-1)}{d}\dot{\bar{K}} - \tfrac{d-1}{d}\bar{K}^2 \right] \psi$$
$$- \tfrac{1}{2}(d-1) \psi \left[\Box_0 - 2\Lambda_k - \tfrac{2(d-1)}{d}\dot{\bar{K}} - \tfrac{d-1}{d}\bar{K}^2 \right] E \Bigg\}. \tag{4.44}$$

Here the operators $\Box_i$ are those defined in (4.27) and the diagonal terms in field space have been simplified by partial integration. Setting $\bar{K} = 0$, the matrix elements resulting from this expression are shown in the third column of Table 4.2.

The ghost action exponentiating the Faddeev-Popov determinant is obtained from the variations (4.8) by evaluating (A.17). The ghost sector then comprises one scalar ghost $\{\bar{c}, c\}$ and one spatial vector ghost $\{\bar{b}_i, b_i\}$ arising from the transformation of F and F_i, respectively. Restricting to terms quadratic in the fluctuation field and choosing $\epsilon_1 = \epsilon_2 = -1$, the result is given by

$$
\begin{aligned}
\Gamma_k^{\text{ghost}} = \int d\tau d^d y \sqrt{\bar{\sigma}} \Big\{ &+\bar{c} \left[\Box_0 + \tfrac{2}{d}\bar{K}\partial_\tau + \dot{\bar{K}} \right] c \\
&+ \bar{b}^i \left[\Box_1 + \tfrac{2}{d}\bar{K}\partial_\tau + \tfrac{1}{d}\dot{\bar{K}} + \tfrac{d-4}{d^2}\bar{K}^2 \right] b_i \Big\}.
\end{aligned}
\tag{4.45}
$$

Notably, the ghost action does not contain a scale-dependent coupling. The results (4.44) and (4.45) then complete the construction of the Hessian $\Gamma_k^{(2)}$.

4.2.4 Regulator and Beta Functions

The Hessians arising from (4.44) and (4.45) contain D-covariant Laplace-type operators only and can thus be evaluated using the standard heat-kernel techniques introduced in Sect. 3.2.3. In particular we resort to a Type I regulator [30], implicitly defined in Eq. (3.29), and we choose the scalar function R_k to be of Litim-form [31]

$$
R_k = (k^2 - \Box_s)\, \theta(k^2 - \Box_s)
\tag{4.46}
$$

At this point the flow of Newton's constant and the cosmological constant has to be expressed in terms of the dimensionless quantities (3.44) and (3.45), whose expression for a $D = (d + 1)$-dimensional manifold is

$$
\eta \equiv (G_k)^{-1} k \partial_k G_k, \qquad \lambda_k \equiv \Lambda_k k^{-2}, \qquad g_k \equiv G_k k^{d-1}.
\tag{4.47}
$$

The scale-dependence of g_k and λ_k is then encoded in

$$
k\partial_k g_k = \beta_g(g, \lambda; d), \qquad k\partial_k \lambda_k = \beta_\lambda(g, \lambda; d).
\tag{4.48}
$$

The explicit expression for the β-functions, whose derivation is reported in Appendix A.2, is the following

$$
\beta_g = (d - 1 + \eta)\, g\,,
$$

$$\beta_\lambda = (\eta - 2)\lambda + \frac{2g}{(4\pi)^{(d-1)/2}} \frac{1}{\Gamma((d+3)/2)} \left[\left(d + \frac{d^2+d-4}{2(1-2\lambda)} + \frac{3d-3-(4d-2)\lambda}{B_{\det}(\lambda)} \right) \left(1 - \frac{\eta}{d+3} \right) \right.$$
$$\left. - 2(d+1) + N_S + (d-1)N_V - 2^{[(d+1)/2]} N_D \right]$$

$$(4.49)$$

where the function $B_{\det}(\lambda)$ is defined as

$$B_{\det}(\lambda) \equiv (1 - 2\lambda)(d - 1 - d\lambda) \tag{4.50}$$

and the anomalous dimension of Newton's constant is given by

$$\eta = \frac{16\pi g B_1(\lambda)}{(4\pi)^{(d+1)/2} + 16\pi g B_2(\lambda)}. \tag{4.51}$$

The functions $B_1(\lambda)$ and $B_2(\lambda)$ depend on λ and d, and read

$$B_1(\lambda) = -\frac{d^5+17d^4+41d^3+85d^2+174d-78}{24\,d(d-1)\,\Gamma((d+5)/2)} + \frac{d^4-5d^2+16d+48}{12\,d(d-1)\,(1-2\lambda)\,\Gamma((d+1)/2)}$$
$$- \frac{d^4-15d^2+28d-10}{2d(d-1)\,(1-2\lambda)\,\Gamma((d+3)/2)} + \frac{3d-3-(4d-2)\lambda}{6\,B_{\det}(\lambda)\,\Gamma((d+1)/2)}$$
$$+ \frac{c_{1,0}+c_{1,1}\lambda+c_{1,2}\lambda^2}{4\,d\,B_{\det}(\lambda)^2\,\Gamma((d+3)/2)} + \frac{1}{6\,\Gamma((d+1)/2)} \left[N_S + \frac{d^2-13}{d+1}N_V - \frac{d-2}{d+1}\,2^{[(d+1)/2]}N_D \right]$$

$$(4.52a)$$

$$B_2(\lambda) = +\frac{d^4-10d^3+21d^2+6d+6}{24\,d(d-1)\,\Gamma((d+5)/2)} + \frac{d^4-5d^2+16d+48}{24\,d(d-1)\,(1-2\lambda)\,\Gamma((d+3)/2)}$$
$$- \frac{d^4-15d^2+28d-10}{4\,d(d-1)\,(1-2\lambda)^2\,\Gamma((d+5)/2)} + \frac{3d-3-(4d-2)\lambda}{12\,B_{\det}(\lambda)\,\Gamma((d+3)/2)} + \frac{c_{2,0}+c_{2,1}\lambda+c_{2,2}\lambda^2}{8\,d\,B_{\det}(\lambda)^2\,\Gamma((d+5)/2)}.$$

$$(4.52b)$$

The coefficients $c_{i,j}$ are polynomials in d, and are given by

$$c_{1,0} = -5d^3 + 22d^2 - 24d + 16\,, \quad c_{1,1} = 4\left(d^3 - 10d^2 + 16d - 16\right),$$
$$c_{1,2} = 4\left(d^3 + 6d^2 - 16d + 16\right), \quad c_{2,0} = -5d^3 + 22d^2 - 24d + 16\,, \tag{4.53}$$

together with $c_{1,1} = c_{2,1}$ and $c_{1,2} = c_{2,2}$. Notably B_2 is independent of the matter content of the system, reflecting the fact that the matter sector (4.18) is independent of Newton's constant. The resulting beta functions (4.49) together with the explicit expression for the anomalous dimension of Newton's constant (4.51) constitutes the main result of this section.

4.3   Properties of the Renormalization Group Flow

In this section, we analyze the renormalization group flow resulting from the beta functions (4.49) for a $D = (3 + 1)$-dimensional spacetime. The case of pure gravity, corresponding to setting $N_S = N_V = N_D = 0$, is discussed in Sect. 4.3.1 while the classification of the fixed point structures appearing in general gravity-matter systems is carried out in Sect. 4.3.2.

4.3.1   Pure Gravity

We start our discussion by studying the fixed point structure arising from the beta functions (4.49) in the case of pure gravity. For $d = 3$ spatial dimensions the system under consideration possesses a unique NGFP with positive Newton's constant

$$\text{NGFP:} \qquad g_* = 0.785\,, \qquad \lambda_* = 0.315\,, \qquad g_*\lambda_* = 0.248\,. \qquad (4.54)$$

The NGFP comes with a complex pair of critical exponents

$$\theta_{1,2} = 0.503 \pm 5.377i\,. \qquad (4.55)$$

The positive real part, $\text{Re}(\theta_{1,2}) > 0$, indicates that the NGFP acts as a spiraling UV attractor for the RG trajectories in its vicinity. Notably, this is the same type of UV-attractive spiraling behavior encountered when evaluating the renormalization group flow on foliated spacetimes using the Matsubara formalism [7, 8], and a vast range of studies building on the metric formalism [32–63].

As we have seen in Sect. 3.3, some solutions of the RG equations may be ill defined in the infrared limit. For this reason, it is important to determine a priori the singular loci of the theory space where either β_g or β_λ diverge. For finite values of g and λ these may either be linked to one of the denominators appearing in β_λ becoming zero or to divergences of the anomalous dimension of Newton's constant. Inspecting β_λ, the first case gives rise to two singular lines in the (λ, g)-plane

$$\lambda_1^{\text{sing}} = \tfrac{1}{2}\,, \qquad \lambda_2^{\text{sing}} = \tfrac{d-1}{d}\,. \qquad (4.56)$$

The singular lines $\eta^{\text{sing}}(g, \lambda)$ associated with divergences of the anomalous dimension η are complicated functions of d. For the specific cases $d = 2$ and $d = 3$ the resulting expressions simplify and are given by the parametric curves

$$d = 2: \ \eta^{\text{sing}} : g = -\frac{45\pi(1-2\lambda)^2}{2\,(76\lambda^2 - 296\lambda + 147)}\,,$$
$$d = 3: \ \eta^{\text{sing}} : g = -\frac{144\pi(6\lambda^2 - 7\lambda + 2)^2}{144\lambda^4 - 1884\lambda^3 + 3122\lambda^2 - 1688\lambda + 279}\,. \qquad (4.57)$$

The position of the singular lines (4.56) and (4.57) are illustrated in Fig. 4.1. Focusing to the domain $g \geq 0$, it is interesting to note that the singularities bounding the flow of λ_k for positive values are of different nature in $d = 2$ and $d = 3$. Specifically, in $d = 2$ the domain is bounded to the right by a fixed singularity of β_λ and η remains finite throughout this domain, while in $d = 3$ the singular line λ_1^{sing} is screened by a

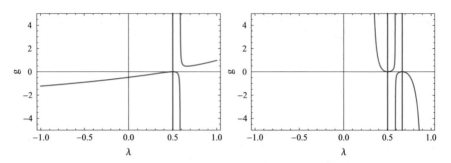

Fig. 4.1 Singularity structure of the beta functions (4.49) in the (λ, g)-plane in $d = 2$ (left diagram) and $d = 3$ (right diagram). The blue lines indicate the singularities of β_λ, Eq. (4.56), while the red lines illustrate the curves (4.57) where η develops a singularity

divergence of η. Notably, the position of the singular lines is independent of N_S, N_V, and N_D and thus also carries over to the analysis of gravity-matter systems. Finally, we note that the point $(\lambda, g) = (1/2, 0)$ is special in the sense that the beta functions (4.49) are of the form 0/0. Notably, in some cases this "quasi-fixed point" $C \equiv (\frac{1}{2}, 0)$ may provide an infrared completion for the gravitational renormalization group flow [29], and we will discuss this possibility in Sect. 4.4. Upon determining the fixed point and singularity structure relevant for the renormalization group flow with a positive Newton's constant, it is rather straightforward to construct the RG trajectories resulting from the beta functions (4.49) numerically. An illustrative sample of RG trajectories characterizing the flow in $D = 3 + 1$ spacetime dimensions is shown in Fig. 4.2. Notably, the high-energy behavior of the flow is controlled by the NGFP (4.54). Following the nomenclature introduced in [34], and summarized in Sect. 3.3, the low-energy behavior of the flow can be classified according to the sign of the cosmological constant. The trajectories belonging to the renormalization group flow can thus be divided into three families: the RG trajectories flowing to the left (Type Ia, $\Lambda_0 < 0$), to the right (Type IIIa, $\Lambda_t > 0$), and that flowing towards the GFP "O" (Type IIa, $\Lambda_0 = 0$), represented by a bold blue line. Once the trajectories enter the vicinity of the GFP, characterized by $g_k \ll 1$, the dimensionful Newton's constant G_k and cosmological constant Λ_k are essentially k-independent, so that the trajectories enter into the "classical regime". For trajectories of Type Ia this regime extends to $k = 0$, while trajectories of Type IIIa terminate in the singularity η^{sing} (red line) at a finite value of the RG scale, $k = k_t$.

The high-energy and low-energy regimes are connected by a crossover of the renormalization group flow. For some of the trajectories, this crossover cuts through the red line marking a divergence in the anomalous dimension of Newton's constant. This peculiar feature can be traced back to the critical exponents (4.55) where the beta functions (4.49) lead to an exceptionally low value for $\text{Re}(\theta_{1,2})$. Compared to other incarnations of the flow, which come with significantly higher values for $\text{Re}(\theta_{1,2})$, this makes the spiraling process around the NGFP less compact. As a consequence the flow actually touches η^{sing}. Since this feature is absent in the flow diagrams obtained

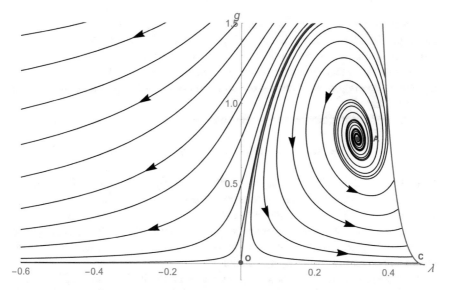

Fig. 4.2 Phase diagram of the renormalization group flow originating from the beta functions (4.49) in $D = 3 + 1$ spacetime dimensions. The flow is dominated by the interplay of the NGFP (point "A") controlling the flow for ultra-high energies and the GFP (point "O") governing the low-energy behavior. The flow undergoes a crossover between these two fixed points. For some of the RG trajectory this crossover is intersected by the singular locus (4.57) (red line). The arrows indicate the direction of the renormalization group flow pointing from high to low energy

from the Matsubara computation [7, 8], the foliated renormalization group flows studied in [29], and the one obtained in the covariant formalism [34], it is likely that this is rather a particularity of the flow based on (4.49), instead of a genuine physical feature.

4.3.2 Gravity-Matter Systems

In this section we study the fixed point structure entailed by (4.49) in the presence of free matter fields. In particular we will specialize to the case of $d = 3$ space dimensions.

In order to classify the fixed point structures realized for a generic gravity-matter system, we first observe that the number of minimally coupled scalar fields, N_S, vectors, N_V, and Dirac spinors N_D enter the gravitational beta functions (4.49) in terms of the combinations

$$d_g \equiv N_S + \frac{d^2 - 13}{d + 1} N_V - 2^{\frac{(d+1)}{2}} \frac{d - 2}{d + 1} N_D, \tag{4.58a}$$

$$d_\lambda \equiv N_S + (d - 1) N_V - 2^{\frac{(d+1)}{2}} N_D. \tag{4.58b}$$

The precise relation between the parameters d_g, d_λ and the matter content may depend on the precise choice of regulator employed in matter traces (see Appendix A.2.3). Carrying out the classification of fixed point structures in terms of the deformation parameters shifts this regulator dependence into the map $(d_g(N_S, N_V, N_D), d_\lambda(N_S, N_V, N_D))$ allowing to carry out the classification independently of a particular regularization scheme.

For $d = 3$ the definitions (4.58) reduce to

$$d_g = N_S - N_V - N_D, \qquad d_\lambda = N_S + 2N_V - 4N_D. \tag{4.59}$$

The relation (4.59) allows to assign coordinates to any matter sector. For example, the Standard Model of particle physics comprises $N_S = 4$ scalars, $N_D = 45/2$ Dirac fermions and $N_V = 12$ vector fields and is thus located at $(d_g, d_\lambda) = (-61/2, -62)$. For N_S and N_V being positive integers including zero and N_D taking half-integer values in order to also accommodate chiral fermions, d_g and d_λ take half-integer values and cover the entire (d_g, d_λ)-plane.

The beta functions (4.49) then give rise to a surprisingly rich set of NGFPs whose properties can partially be understood analytically. The condition $\beta_g|_{g=g_*} = 0$ entails that any NGFP has to come with an anomalous dimension $\eta_* = -2$. This relation can be solved analytically, determining the fixed point coordinate $g_*(\lambda_*; d_g)$ as a function of λ_* and d_g. Substituting $\eta_* = -2$ together with the relation for g_* into the second fixed point condition, $\beta_\lambda|_{g=g_*} = 0$, then leads to a fifth order polynomial in λ whose coefficients depend on d_g, d_λ. The roots of this polynomial provide the coordinate λ_* of a candidate NGFP. The fact that the polynomial is of fifth order then implies that the beta functions (4.49) may support at most five NGFPs, independently of the matter content of the system.

The precise fixed point structure realized for a particular set of values (d_g, d_λ) can be determined numerically. The number of NGFPs located within the physically interesting region $g_* > 0$ and $\lambda_* < 1/2$ is displayed in Fig. 4.3, where black, blue, green and red mark matter sectors giving rise to zero, one, two, and three NGFPs, respectively. On this basis, we learn that systems possessing zero or one NGFP are rather generic, while matter sectors giving rise to two or three NGFPs are confined to a small region in the center of the (d_g, d_λ)-plane.

The classification of the NGFPs according to their stability properties is provided in Fig. 4.4 with the color-coding explained in Table 4.3. The left diagram provides the classification for the case of zero (black region) and one NGFP. Here green and blue indicate the existence of a single UV-attractive NGFP with real (green) or complex (blue) critical exponents. Saddle points with one UV attractive and one UV repulsive eigendirections (magenta) and IR fixed points (red, orange) occur along a small wedge paralleling the d_λ-axis, only. The gray region supporting multiple NGFPs is magnified in the right diagram of Fig. 4.3. All points in this region support at least one UV NGFP suitable for Asymptotic Safety while there is a wide range of possibilities for the stability properties of the second and third NGFP. The classification in Fig. 4.3 establishes that the existence of a UV-attractive NGFP suitable for Asymptotic Safety

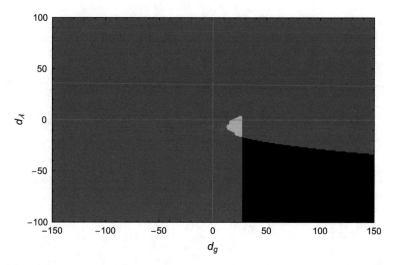

Fig. 4.3 Number of NGFPs supported by the beta functions (4.49) as a function of the parameters d_g and d_λ. The colors black, blue, green, and red indicate the existence of zero, one, two, and three NGFPs situated at $g_* > 0$, $\lambda_* < 1/2$, respectively

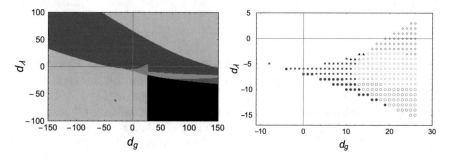

Fig. 4.4 Classification of the NGFPs arising from the beta functions (4.49) in the (d_g, d_λ)-plane, following the color-code provided in Table 4.3. The left diagram classifies the stability behavior of the one-fixed point sector. In particular, the black region does not support any NGFP while the regions giving rise to a single, UV-attractive NGFP with complex and real critical exponents are marked in blue and green, respectively. The field content of the Standard Model is situated in the lower-left quadrant, $(d_g, d_\lambda) = (-61/2, -62)$, and marked with a bold dot. The gray area, supporting multiple NGFPs is magnified in the right diagram with empty and filled symbols indicating the existence of two and three NGFPs, respectively

is rather generic and puts only mild constraints on the admissible values (d_g, d_λ). At this stage, it is interesting to relate this classification to phenomenologically interesting matter sectors including the Standard Model of particle physics (SM) and its most commonly studied extensions.[3] The result is summarized in Table 4.4. The map (4.59) allows to relate the number of scalars N_S, vector fields N_V and Dirac

---

[3]For a similar discussion within metric approach to Asymptotic Safety see [24].

Table 4.3 Color-code for the fixed point classification provided in Fig. 4.4. The column NGFPs gives the number of NGFP solutions while the subsequent columns characterize their behavior in terms of 2 UV-attractive (UV), one UV-attractive and one UV-repulsive (saddle) and 2 IR-attractive (IR) eigendirections with real (real) and complex (spiral) critical exponents

Class	NGFPs	NGFP$_1$	NGFP$_2$	NGFP$_3$	Color code
Class 0	0	–	–	–	Black region
Class Ia	1	UV, spiral	–	–	Blue region
Class Ib	1	UV, real	–	–	Green region
Class Ic	1	Saddle	–	–	Magenta region
Class Id	1	IR, spiral	–	–	Red region
Class Ie	1	IR, real	–	–	Orange region
Class IIa	2	UV, real	IR, real	–	Open circle
Class IIb	2	UV, real	IR, spiral	–	Open square
Class IIc	2	UV, spiral	IR, spiral	–	Open triangle
Class IId	2	UV, spiral	UV, real	–	Open diamond
Class IIIa	3	UV, real	Saddle	IR, real	Filled circle
Class IIIb	3	UV, real	Saddle	IR, spiral	Filled square
Class IIIc	3	UV, spiral	Saddle	IR, spiral	Filled triangle

fermions N_D defining the field content of a specific matter sector to coordinates in the (d_g, d_λ)-plane. The resulting coordinates are given in the fifth and sixth column of Table 4.4. Correlating these coordinates with the data provided by Fig. 4.4 yields two important results. Firstly, all matter models studied in Table 4.4 are located in regions of the (d_g, d_λ)-plane which host a single UV-attractive NGFP with real stability coefficients. Secondly, we note a qualitative difference between the Standard Model (SM) and its extensions (first five matter sectors) and Grand Unified Theories (GUTs). The former all belong to the green region in the lower left part of the (d_g, d_λ)-plane while the second class of models sits in the upper-right quadrant. As a result, the corresponding NGFPs possess very distinct features. The NGFPs appearing in the first case have a characteristic product $g_* \lambda_* < 0$. Their critical exponents show a rather minor dependence on the precise matter content of the theory and have values in the range $\theta_1 \sim 3.8 - 4.0$ and $\theta_2 \sim 2.0$. In contrast, the NGFPs appearing in the context of GUT-type models come with a positive product $g_* \lambda_* > 0$. Their critical exponents are significantly larger $\theta_1 > 19$ than in the former case and show a much stronger dependence on the matter field content. Thus while all matter sectors investigated in Table 4.4 give rise to a NGFP suitable for realizing Asymptotic Safety the magnitude of the critical exponents hints that the SM-type theories may have more predictive power in terms of a lower number of relevant coupling constants in the gravitational sector.

At this stage it is also instructive to construct the phase diagram resulting from gravity coupled to the matter content of the Standard Model. An illustrative sample

Table 4.4 Fixed point structure arising from the field content of commonly studied matter models. All models apart from the Minimal Supersymmetric Standard Model (MSSM) and the Grand Unified Theory (GUT), sit in the lower-left quadrant of Fig. 4.4. All matter configurations possess a single ultraviolet attractive NGFP with real critical exponents

Model	N_S	N_D	N_V	d_g	d_λ	g_*	λ_*	θ_1	θ_2
Pure gravity	0	0	0	0	0	0.78	$+0.32$	$0.50 \pm 5.38\,i$	
Standard Model (SM)	4	$\frac{45}{2}$	12	$-\frac{61}{2}$	-62	0.75	-0.93	3.871	2.057
SM, dark matter (dm)	5	$\frac{45}{2}$	12	$-\frac{59}{2}$	-61	0.76	-0.94	3.869	2.058
SM, $3\,\nu$	4	24	12	-32	-68	0.72	-0.99	3.884	2.057
SM, $3\,\nu$, dm, axion	6	24	12	-30	-66	0.75	-1.00	3.882	2.059
MSSM	49	$\frac{61}{2}$	12	$+\frac{13}{2}$	-49	2.26	-2.30	3.911	2.154
SU(5) GUT	124	24	24	$+76$	$+76$	0.17	$+0.41$	25.26	6.008
SO(10) GUT	97	24	45	$+28$	$+91$	0.15	$+0.40$	19.20	6.010

of RG trajectories obtained from solving the beta functions (4.49) for $(d_g, d_\lambda) = (-61/2, -62)$ is shown in Fig. 4.5. Similarly to the case of pure gravity, the flow is dominated by the interplay of the NGFP situated at $(g_*, \lambda_*) = (0.75, -0.93)$ and the GFP in the origin. The NGFP controls the UV behavior of the trajectories while the GFP is responsible for the occurrence of a classical low-energy regime. The classification of possible low-energy behaviors is again given by the limits (3.56). A notable difference to the pure gravity case is the absence of the spiraling behavior of trajectories onto the NGFP. This reflects the property that the NGFPs of the gravity-matter models come with real critical exponents. Moreover, the shift of the NGFP to negative values λ_* entails that the singularity (4.57) (red line) no longer affects the crossover of the trajectories from the NGFP to the GFP. Other matter sectors located in the lower-left green region of Fig. 4.4 give rise to qualitatively similar phase diagrams so that the flow shown in Fig. 4.5 provides a prototypical showcase for this class of universal behaviors. Notably, these peculiar features of gravity-matter systems will be fundamental for the discussion of Chap. 5.

Owed to their relevance for cosmological model building, we close this section with a more detailed investigation of the fixed point structures appearing in gravity-scalar models with $N_V = N_D = 0$. For illustrative purposes we formally also include negative values N_S in order to capture the typical behavior of matter theories located in the lower-left quadrant of Fig. 4.4. Notably, all values N_S give rise to a NGFP with two UV attractive eigendirections. The position (λ_*, g_*) and stability coefficients of this family of fixed points is displayed in Fig. 4.6. The first noticeable feature is a sharp transition in the position of the NGFP occurring at $N_S \sim -6$: for $N_S \le -6$ the NGFP is located at $\lambda_* < 0$ while for $N_S > -5$ one has $\lambda_* > 0$. For $N_S \to \infty$ the fixed point approaches $C \equiv (1/2, 0)$, which can be shown to be a fixed point of the beta functions (4.49) in the large N_S limit. The value of the critical exponents shown in the lower line of Fig. 4.6 indicates that there are two transitions: for $N_S \le -6$ there is a UV-attractive NGFP with two real critical exponents $\theta_1 \sim 4$ and $\theta_2 \sim 2$. These values are essentially independent on N_S. On the interval $-6 \le N_S \le 46$ the

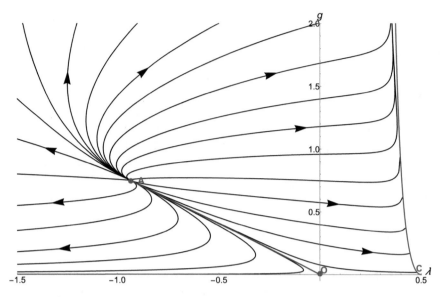

Fig. 4.5 Phase diagram depicting the renormalization group flow of gravity coupled to the matter content of the Standard Model in $D = 3 + 1$ spacetime dimensions. Similarly to the pure gravity case, the phase diagram is determined by the interplay of the NGFP (point "A") controlling the flow for ultra-high energies and the GFP (point "O") governing its low-energy behavior. The singular locus (4.57) is depicted by the red line and arrows point towards lower values of k

critical exponents turn into a complex pair. In particular for $N_S = 0$, one recovers the pure gravity fixed point NGFP (4.54). For $N_S > 46$ one again has a UV-attractive NGFP with two real critical exponents with one of the critical exponents becoming large. Thus we clearly see a qualitatively different behavior of the NGFPs situated in the upper-right quadrant (relevant for GUT-type matter models) and the NGFPs in the lower-left quadrant (relevant for the Standard Model) of Fig. 4.4, reconfirming that the Asymptotic Safety mechanism realized within these classes of models is of a different nature.

4.4 Universality Classes for Quantum Gravity

The Wilsonian Renormalization Group encodes the universal critical behavior of a physical system through the fixed points of the renormalization group flow. Within the context of Quantum Gravity, the ADM-formalism allows to study such a critical behavior in the presence of a foliation structure, and the beta functions generally depend on the spacetime dimension $D = d + 1$ and the matter content of the theory. In the previous section we studied the fixed point structure arising from the beta functions (4.49) in the presence of an arbitrary number of matter fields living in a $(3 + 1)$-dimensional spacetime. In this section, we discuss the fixed point structure

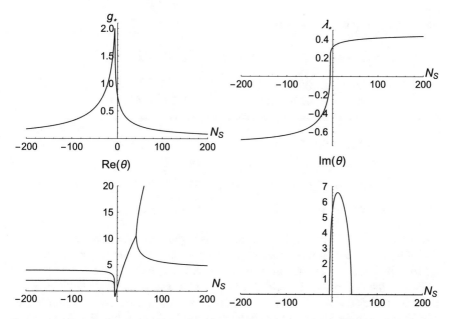

Fig. 4.6 Position (top) and stability coefficients (bottom) of the UV NGFPs appearing in gravity-scalar systems as a function of N_S. The fixed point structure undergoes qualitative changes at $N_S \sim -6$ and $N_S \sim 46$ where the critical exponents change from real to complex values

emerging from the renormalization group flow of pure foliated Quantum Gravity as a function of the spacetime dimension [29].

4.4.1 Fixed Points and Universality Classes

In the present discussion we use the beta functions derived in [29] for a $(d + 1)$-dimensional spacetime. The latter differ from the ones used in [27], Eq. (4.49), by a different form of the regulator in the transverse-traceless and vector sectors of the decompositions (4.34) and (4.35). The resulting beta functions (β_g, β_λ) and anomalous dimension η still have the form (4.49) and (4.51) respectively, while the functions $B_1(\lambda)$ and $B_2(\lambda)$ defining the anomalous dimension (4.51) acquire new contributions. In particular they read

$$B_1(\lambda) = -\frac{d^4+14d^3-d^2+94d+12}{12\,d\,(d-1)\,\Gamma((d+3)/2)} + \frac{d^2+d-4}{12\,(1-2\lambda)\,\Gamma((d+1)/2)} - \frac{d^4-15d^2+28d-10}{2d\,(d-1)\,(1-2\lambda)^2\,\Gamma((d+3)/2)}$$
$$+ \frac{3d-3-(4d-2)\lambda}{6\,B_{\mathrm{det}}(\lambda)\,\Gamma((d+1)/2)} + \frac{c_{1,0}+c_{1,1}\lambda+c_{1,2}\lambda^2+c_{1,3}\lambda^3+c_{1,4}\lambda^4}{4\,d\,(d^2+2d-3)\,B_{\mathrm{det}}(\lambda)^2\,\Gamma((d+3)/2)} \ , \tag{4.60a}$$

$$B_2(\lambda) = +\frac{d^3-9d^2+12d+12}{24\,d\,\Gamma((d+5)/2)} + \frac{d^2+d-4}{24\,(1-2\lambda)\,\Gamma((d+3)/2)} - \frac{d^4-15d^2+28d-10}{4\,d\,(d-1)\,(1-2\lambda)^2\,\Gamma((d+5)/2)}$$
$$+ \frac{3d-3-(4d-2)\lambda}{12\,B_{\mathrm{det}}(\lambda)\,\Gamma((d+3)/2)} + \frac{c_{2,0}+c_{2,1}\lambda+c_{2,2}\lambda^2}{8\,d\,B_{\mathrm{det}}(\lambda)^2\,\Gamma((d+5)/2)} \ , \tag{4.60b}$$

with the coefficients $c_{i,j}$ given by

$$
\begin{aligned}
c_{1,0} &= -(d-1)(5d^4 - 7d^3 - 74d^2 + 56d - 16)\,, \\
c_{1,1} &= +4(d-1)(d^4 - 7d^3 - 62d^2 + 16d - 16)\,, \\
c_{1,2} &= +4d^5 + 32d^4 + 388d^3 - 232d^2 - 64d - 64\,, \\
c_{1,3} &= -128\,d\,(d+1)(3d-2)\,, \\
c_{1,4} &= +128\,d^2\,(d+1)\,, \\
c_{2,0} &= -5d^3 + 22d^2 - 24d + 16\,, \\
c_{2,1} &= +4d^3 - 40d^2 + 64d - 64\,, \\
c_{2,2} &= +4d^3 + 24d^2 - 64d + 64\,.
\end{aligned}
\tag{4.61}
$$

The resulting $(d+1)$-dimensional β functions together with the anomalous dimension of Newton's constant completely encode the renormalization group flow resulting from the ansatz (4.11) and give rise to a rich fixed point structure.

Firstly, there is the GFP located in the origin, whose stability coefficients are given by the classical scaling dimensions. In addition, there are two families of NGFPs whose most important properties are summarized in Fig. 4.7. In $D = 2 + \varepsilon$ dimensions a NGFP with two real, positive stability coefficients (orange line) emerges from the GFP. Its critical exponents agree with the epsilon-expansion of perturbative gravity around two dimensions [64] to leading order. This universality class has an upper critical dimension $D = 2.28$. At this point there is a transition to a family of saddle points (SP) characterized by a small UV-attractive and a large UV-repulsive critical exponent (brown line). In $D = 2.37$ these critical exponents swap sign, giving rise to the red line of SP-NGFPs existing for $2.37 \leq D \leq 3.25$. Simultaneously, there is a second family of fixed points (green line) with two real, positive critical exponents. At $D = 3.40$ the two stability coefficients coincide at $\theta_1 = \theta_2 = 1.08$. For $D > 3.40$ the real stability coefficients become complex (blue line) which reflects the typical characteristics of the UV-NGFP seen in the metric approach to Asymptotic Safety. The additional information displayed in Fig. 4.7 encodes the stability coefficients found within related Quantum Gravity programs. The qualitative agreement of the scaling behavior seen within discrete (Monte Carlo) methods in $D = 2 + 1$ dimensions and the continuum RG is an important indicator for the robustness of the underlying universality classes.

The properties of the NGFPs in $D = 2 + 1$ and $D = 3 + 1$ are summarized as follows. In $D = 2 + 1$, the UV-NGFP and SP-NGFPs are located at

$$
\begin{array}{llll}
\text{UV-NGFP:} & g_* = 0.16\,, & \lambda_* = 0.03\,, & g_*\lambda_* = 0.005\,, \\
\text{SP-NGFP:} & g_* = 0.32\,, & \lambda_* = 0.20\,, & g_*\lambda_* = 0.07\,,
\end{array}
\tag{4.62}
$$

and come with stability coefficients

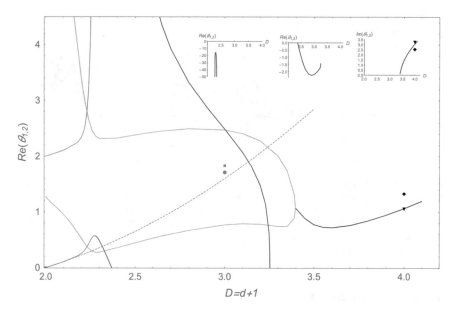

Fig. 4.7 Stability coefficients of the two families of fixed points emerging from the β functions as a function of the spacetime dimension $D = d + 1$. The dashed line gives the result from the two-loop epsilon expansion [64], the circles indicate the scaling of Newton's coupling found within lattice Quantum Gravity [65] and the square marks the scaling found from the exact solution of the discretized Wheeler-de Witt equation [66, 67]. In $D = 4$ the down-triangle indicates the critical exponents obtained from foliated spacetimes using the Matsubara formalism [7] while the diamond corresponds to the dynamical fixed point seen in [51]

$$
\begin{array}{llll}
\text{UV-NGFP:} & \theta_1 = +2.47, & \theta_2 = +0.77, \\
\text{SP-NGFP:} & \theta_1 = +2.49, & \theta_2 = -2.20.
\end{array}
\tag{4.63}
$$

In $D = 3 + 1$ there is a unique NGFP for positive Newton's constant

$$
\text{NGFP:} \quad g_* = 0.90, \quad \lambda_* = 0.24, \quad g_*\lambda_* = 0.21,
\tag{4.64}
$$

with critical exponents

$$
\text{NGFP:} \quad \theta_{1,2} = 1.06 \pm 3.07i.
\tag{4.65}
$$

This NGFP exhibits the typical complex pair of critical exponents familiar from evaluating the flow equation in the metric formulation [32–38, 51, 56, 57]. In particular, there is a very good agreement with the critical exponents obtained for foliated spacetime via the Matsubara formalism [7, 8]. Thus it is highly conceivable that the NGFPs seen in these computations belong to the same universality class.

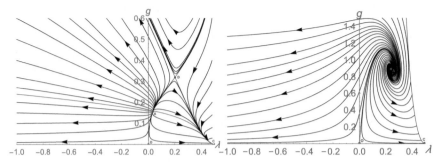

Fig. 4.8 Phase diagrams resulting from the β functions of [29] for $D = 2 + 1$ (left) and $D = 3 + 1$ (right) spacetime dimensions. The GFP, NGFP and SP-NGFP are marked by the points "O", "A", and "D". In $D = 2 + 1$ the interplay of the fixed points "A" and "D" ensures that the QFP "C" provides the long-distance completion of the renormalization group trajectories with a positive cosmological constant. In $D = 3 + 1$ the SP-NGFP "D" is absent and the corresponding trajectories terminate in a divergence of η (red line)

4.4.2   Phase Diagrams

The phase diagram resulting from the numerical integration of the β functions of [29] in $D = 2 + 1$ and $D = 3 + 1$ is shown in the left and right panel of Fig. 4.8. For $D = 2 + 1$ dimensions, the flow is governed by the interplay of the GFP "O", the two NGFPs and the two quasi fixed points (QFPs) "B" and "C" located at $(\lambda, g) = (-\infty, 0)$ and $(\lambda, g) = (1/2, 0)$, respectively. Notably, the point "C" is the same QFP encountered in Sect. 4.3.1. The separatrix lines connecting these fixed points are depicted in blue. In particular, all RG trajectories located below the line $\overline{ODC}$ are well-behaved for all $k \in [0, \infty[$ and their high-energy behavior is controlled by the UV-NGFP "A". Lowering the RG scale they cross over to the GFP or the saddle point "D". Trajectories passing sufficiently close to the GFP develop a classical regime where both the Newton's coupling and cosmological constant are independent of the RG scale. Finally, depending on whether the flow approaches "B", "O", or "C" in the infrared limit, the classical regime exhibits a negative, zero, or positive cosmological constant, thus realizing the same classification described in Sect. 3.3.

The picture found in $D = 3 + 1$ dimensions is similar: the UV completion of the flow is controlled by the NGFP and the classical regime emerges from the crossover of the flow to the GFP. As a consequence of the missing SP-NGFP "D" the trajectory $\overline{CD}$ in $D = 2 + 1$ is replaced by a line of singularities where η diverges (red line). In this case solutions with a positive cosmological constant terminate at a finite RG scale, k_t. Note that this phase diagram has the same properties of that shown in Fig. 4.2.

Since the RG trajectory describing our world should exhibit a classical regime with a positive cosmological constant [68], it is worthwhile to investigate the mechanism providing the IR completion of these trajectories in detail. Figure 4.9 displays the typical scale-dependence of the Newton's coupling (left) and the cosmological

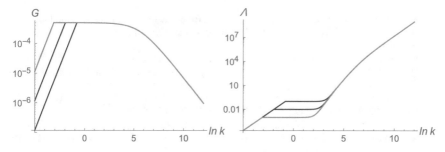

Fig. 4.9 Scale dependence of Newton's constant G_k (left) and the cosmological constant (right) for a set of RG trajectories exhibiting a classical regime with a positive cosmological constant for $D = 2 + 1$. The QFP "C" drives the infrared-value of Λ to zero dynamically

constant (right) for this class of solutions. Starting from the high-energy part and decreasing the RG scale k, the trajectories undergo four phases. The fixed point regime is controlled by the UV-NGFP ensuring the absence of unphysical UV divergences. Subsequently, there is a semi-classical regime followed by a classical regime where the Newton's constant and cosmological constant exhibit plateaus. In particular the latter phases are independent of the spacetime dimension. In $D = 2 + 1$ the low-energy completion of the solutions is provided by a novel phase where both G and Λ are dynamically driven to zero. This phase is controlled by the QFP "C" situated at $(\lambda, g) = (1/2, 0)$. At this point the β functions (4.49) are ambiguous owed to terms of the form $g/(1 - 2\lambda)^2$ where both the numerator and denominator vanish. RG trajectories approaching "C" resolve this ambiguity in such a way that $\lim_{k \to 0} g/(1 - 2\lambda)^2 = 5\pi/6$. In this way "C" is turned into a *low-energy attractor* where $\eta = 2$. The mechanism providing the low-energy completion in this sector is essentially the same as the one reported in [47]. Notably, if this mechanism worked in four spacetime dimensions, the effects of such a gravitational phase (related to the physics of a QFP similar to "C") would be visible on cosmic scales.

References

1. R. Arnowitt, S. Deser, C.W. Misner, Dynamical structure and definition of energy in general relativity. Phys. Rev. **116**, 1322–1330 (1959). https://doi.org/10.1103/PhysRev.116.1322 (cit. on p. 59)
2. R. Arnowitt, S. Deser, C.W. Misner, Republication of: the dynamics of general relativity. Gen. Relativ. Gravit. **40**, 1997–2027 (2008). https://doi.org/10.1007/s10714-008-0661-1. (cit. on p. 59)
3. C. Wetterich, Exact evolution equation for the effective potential. Phys. Lett. B **301**, 90–94 (1993). https://doi.org/10.1016/0370-2693(93)90726-X (cit. on p. 59)
4. T.R. Morris, The exact renormalization group and approximate solutions. Int. J. Mod. Phys. A **9**, 2411–2449 (1994). https://doi.org/10.1142/S0217751X94000972. eprint: hep-ph/9308265 (cit. on p. 59)

5. M. Reuter, C. Wetterich, Effective average action for gauge theories and exact evolution equations. Nucl. Phys. B **417**, 181–214 (1994). https://doi.org/10.1016/0550-3213(94)90543-6 (cit. on p. 59)

6. M. Reuter, Nonperturbative evolution equation for quantum gravity. Phys. Rev. D **57**, 971–985 (1998). https://doi.org/10.1103/PhysRevD.57.971. eprint: hep-th/9605030 (cit. on pp. 59, 63)

7. E. Manrique, S.Rechenberger, F. Saueressig, Asymptotically safe Lorentzian gravity. Phys. Rev. Lett. **106**(25), 251302 (2011). https://doi.org/10.1103/PhysRevLett.106.251302. arXiv:1102.5012 [hep-th] (cit. on pp. 59, 60, 63, 65, 74, 77, 85, 86)

8. S. Rechenberger, F. Saueressig, A functional renormalization group equation for foliated spacetimes. J. High Energy Phys. **3**, 10 (2013). https://doi.org/10.1007/JHEP03(2013)010. arXiv:1212.5114 [hep-th] (cit. on pp. 59, 60, 63–65, 74, 77, 86)

9. R. Percacci, D. Perini, Constraints on matter from asymptotic safety. Phys. Rev. D **67**(8), 081503 (2003). https://doi.org/10.1103/PhysRevD.67.081503. eprint: hep-th/0207033 (cit. on pp. 60, 64)

10. R. Percacci, D. Perini, Asymptotic safety of gravity coupled to matter. Phys. Rev. D **68**(4), 044018 (2003). https://doi.org/10.1103/PhysRevD.68.044018. eprint: hep-th/0304222 (cit. on pp. 60, 64)

11. O. Zanusso et al., Gravitational corrections to Yukawa systems. Phys. Lett. B **689**, 90–94 (2010). https://doi.org/10.1016/j.physletb.2010.04.043. arXiv:0904.0938 [hep-th] (cit. on p. 60)

12. G.P. Vacca, O. Zanusso, Asymptotic safety in Einstein gravity and scalar-fermion matter. Phys. Rev. Lett. **105**(23), 231601 (2010). https://doi.org/10.1103/PhysRevLett.105.231601. arXiv:1009.1735 [hep-th] (cit. on p. 60)

13. U. Harst, M. Reuter, QED coupled to QEG. J. High Energy Phys. **5**, 119 (2011). https://doi.org/10.1007/JHEP05(2011)119. arXiv:1101.6007 [hep-th] (cit. on p. 60)

14. A. Eichhorn, H. Gies, Light fermions in quantum gravity. New J. Phys. **13**(12), 125012 (2011). https://doi.org/10.1088/1367-2630/13/12/125012. arXiv:1104.5366 [hep-th] (cit. on p. 60)

15. P. Donà, A. Eichhorn, R. Percacci, Matter matters in asymptotically safe quantum gravity. Phys. Rev. D **89**(8), 084035 (2014). https://doi.org/10.1103/PhysRevD.89.084035. arXiv:1311.2898 [hep-th] (cit. on pp. 60, 64, 79)

16. P. Donà, A. Eichhorn, R. Percacci, Consistency of matter models with asymptotically safe quantum gravity. Can. J. Phys. **93**, 988–994 (2015). https://doi.org/10.1139/cjp-2014-0574. arXiv:1410.4411 [gr-qc] (cit. on p. 60)

17. J. Meibohm, J.M. Pawlowski, M. Reichert, Asymptotic safety of gravity-matter systems. Phys. Rev. D **93**(8), 084035 (2016). https://doi.org/10.1103/PhysRevD.93.084035. arXiv:1510.07018 [hep-th] (cit. on pp. 60, 64)

18. K.Y. Oda, M. Yamada, Non-minimal coupling in Higgs-Yukawa model with asymptotically safe gravity. Class. Quantum Gravity **33**(12), 125011 (2016). https://doi.org/10.1088/0264-9381/33/12/125011. arXiv:1510.03734 [hep-th] (cit. on p. 60)

19. P. Donà et al., Asymptotic safety in an interacting system of gravity and scalar matter. Phys. Rev. D **93**(4), 044049 (2016). https://doi.org/10.1103/PhysRevD.93.044049. arXiv:1512.01589 [gr-qc] (cit. on pp. 60, 64)

20. J. Meibohm, J.M. Pawlowski, Chiral fermions in asymptotically safe quantum gravity. Eur. Phys. J. C **76**, 285 (2016). https://doi.org/10.1140/epjc/s10052-016-4132-7. arXiv:1601.04597 [hep-th] (cit. on p. 60)

21. A. Eichhorn, A. Held, J.M. Pawlowski, Quantum-gravity effects on a Higgs-Yukawa model. Phys. Rev. D **94**(10), 104027 (2016). https://doi.org/10.1103/PhysRevD.94.104027. arXiv:1604.02041 [hep-th] (cit. on p. 60)

22. T. Henz, J.M. Pawlowski, C. Wetterich, Scaling solutions for dilaton quantum gravity. Phys. Lett. B **769**, 105–110 (2017). https://doi.org/10.1016/j.physletb.2017.01.057. arXiv:1605.01858 [hep-th] (cit. on p. 60)

23. P. Labus, R. Percacci, G.P. Vacca, Asymptotic safety in O(N) scalar models coupled to gravity. Phys. Lett. B **753**, 274–281 (2016). https://doi.org/10.1016/j.physletb.2015.12.022. arXiv:1505.05393 [hep-th] (cit. on p. 64)

24. P. Donà, R. Percacci, Functional renormalization with fermions and tetrads. Phys. Rev. D **87**(4), 045002 (2013). https://doi.org/10.1103/PhysRevD.87.045002. arXiv:1209.3649 [hep-th] (cit. on p. 64)

25. N. Ohta, R. Percacci, G.P. Vacca, Flow equation for f(R) gravity and some of its exact solutions. Phys. Rev. D **92**(6), 061501 (2015). https://doi.org/10.1103/PhysRevD.92.061501. arXiv:1507.00968 [hep-th] (cit. on p. 64)

26. N. Ohta, R. Percacci, G.P. Vacca, Renormalization group equation and scaling solutions for f(R) gravity in exponential parametrization. Eur. Phys. J. C **76**, 46 (2016). https://doi.org/10. 1140/epjc/s10052-016-3895-1. arXiv:1511.09393 [hep-th] (cit. on p. 64)

27. J. Biemans, A. Platania, F. Saueressig, Renormalization group fixed points of foliated gravity-matter systems. J. High Energy Phys. **5**, 93 (2017). https://doi.org/10.1007/JHEP05(2017)093. arXiv:1702.06539 [hep-th] (cit. on pp. 65, 83)

28. D. Baumann, Inflation, in *Physics of the Large and the Small: TASI 2009*, ed. by C. Csaki, S. Dodelson (2011), pp. 523–686. https://doi.org/10.1142/9789814327183_0010 (cit. on p. 68)

29. J. Biemans, A. Platania, F. Saueressig, Quantum gravity on foliated spacetimes: asymptotically safe and sound. Phys. Rev. D **95**(8), 086013 (2017). https://doi.org/10.1103/PhysRevD.95. 086013. arXiv:1609.04813 [hep-th] (cit. on pp. 71, 75, 77, 83, 86, 87)

30. A. Codello, R. Percacci, C. Rahmede, Investigating the ultraviolet properties of gravity with a Wilsonian renormalization group equation. Ann. Phys. **324**, 414–469 (2009). https://doi.org/ 10.1016/j.aop.2008.08.008. arXiv:0805.2909 [hep-th] (cit. on p. 72)

31. D.F. Litim, Optimized renormalization group flows. Phys. Rev. D **64**(10), 105007 (2001). https://doi.org/10.1103/PhysRevD.64.105007. eprint: hep-th/ 0103195 (cit. on p. 72)

32. W. Souma, Non-trivial ultraviolet fixed point in quantum gravity. Prog. Theor. Phys. **102**, 181–195 (1999). https://doi.org/10.1143/PTP.102.181. eprint: hep-th/9907027 (cit. on pp. 74, 86)

33. O. Lauscher, M. Reuter, Ultraviolet fixed point and generalized flow equation of quantum gravity. Phys. Rev. D **65**(2), 025013 (2002). https://doi.org/10.1103/PhysRevD.65.025013. eprint: hep-th/0108040 (cit. on pp. 74, 86)

34. M. Reuter, F. Saueressig, Renormalization group flow of quantum gravity in the Einstein-Hilbert truncation. Phys. Rev. D **65**(6), 065016 (2002). https://doi.org/10.1103/PhysRevD.65. 065016. eprint: hep-th/0110054 (cit. on pp. 74, 75, 77, 86)

35. D.F. Litim, Fixed points of quantum gravity. Phys. Rev. Lett. **92**(20), 201301 (2004). https:// doi.org/10.1103/PhysRevLett.92.201301. eprint: hep-th/0312114 (cit. on pp. 74, 86)

36. P. Fischer, D.F. Litim, Fixed points of quantum gravity in extra dimensions. Phys. Lett. B **638**, 497–502 (2006). https://doi.org/10.1016/j.physletb.2006.05.073. eprint: hep-th/0602203 (cit. on pp. 74, 86)

37. A. Codello, G. D'Odorico, C. Pagani, Consistent closure of renormalization group flow equations in quantum gravity. Phys. Rev. D **89**(8), 081701 (2014). https://doi.org/10.1103/ PhysRevD.89.081701. arXiv:1304.4777 [gr-qc] (cit. on pp. 74, 86)

38. S. Nagy et al., Critical exponents in Quantum Einstein Gravity. Phys. Rev. D **88**(11), 116010 (2013). https://doi.org/10.1103/PhysRevD.88.116010. arXiv:1307.0765 [hep-th] (cit. on pp. 74, 86)

39. N. Christiansen et al., Local quantum gravity. Phys. Rev. D **92**(12), 121501 (2015). https://doi. org/10.1103/PhysRevD.92.121501. arXiv:1506.07016 [hep-th] (cit. on p. 74)

40. A. Codello, R. Percacci, Fixed points of higher-derivative gravity. Phys. Rev. Lett. **97**(22), 221301 (2006). https://doi.org/10.1103/PhysRevLett.97.221301. eprint: hep-th/0607128 (cit. on p. 74)

41. A. Codello, R. Percacci, C. Rahmede, Ultraviolet properties of f(R)-gravity. Int. J. Mod. Phys. A **23**, 143–150 (2008). https://doi.org/10.1142/S0217751X08038135. arXiv:0705.1769 [hep-th] (cit. on p. 74)

42. D. Benedetti, P.F. Machado, F. Saueressig, Asymptotic safety in higher-derivative gravity. Mod. Phys. Lett. A **24**, 2233–2241 (2009). https://doi.org/10.1142/S0217732309031521. arXiv:0901.2984 [hep-th] (cit. on p. 74)

43. D. Benedetti, P.F. Machado, F. Saueressig, Taming perturbative divergences in asymptotically safe gravity. Nucl. Phys. B **824**, 168–191 (2010). https://doi.org/10.1016/j.nuclphysb.2009.08. 023. arXiv:0902.4630 [hep-th] (cit. on p. 74)

44. M. Demmel, F. Saueressig, O. Zanusso, RG flows of Quantum Einstein Gravity in the linear-geometric approximation. Ann. Phys. **359**, 141–165 (2015). https://doi.org/10.1016/j.aop. 2015.04.018. arXiv:1412.7207 [hep-th] (cit. on p. 74)

45. P.F. Machado, F. Saueressig, On the renormalization group flow of f(R)-gravity. Phys. Rev. D **77**, 124045 (2008). https://doi.org/10.1103/PhysRevD.77.124045. eprint: arXiv:0712.0445 (cit. on p. 74)

46. E. Manrique, M. Reuter, F. Saueressig, Bimetric renormalization group flows in Quantum Einstein Gravity. Ann. Phys. **326**, 463–485 (2011). https://doi.org/10.1016/j.aop.2010.11.006. arXiv:1006.0099 [hep-th] (cit. on p. 74)

47. N. Christiansen et al., Fixed points and infrared completion of quantum gravity. Phys. Lett. B **728**, 114–117 (2014). https://doi.org/10.1016/j.physletb.2013.11.025 (cit. on pp. 74, 88)

48. N. Christiansen et al., Global flows in quantum gravity. Phys. Rev. D **93**(4), 044036 (2016). https://doi.org/10.1103/PhysRevD.93.044036 (cit. on p. 74)

49. K. Falls et al., Further evidence for asymptotic safety of quantum gravity. Phys. Rev. D **93**(10), 104022 (2016). https://doi.org/10.1103/PhysRevD.93.104022 (cit. on p. 74)

50. K. Falls, Critical scaling in quantum gravity from the renormalisation group (2015). arXiv:1503.06233 [hep-th] (cit. on p. 74)

51. I. Donkin, J.M. Pawlowski, The phase diagram of quantum gravity from diffeomorphism-invariant RG-flows (2012). arXiv:1203.4207 [hep-th] (cit. on pp. 74, 85, 86)

52. O. Lauscher, M. Reuter, Flow equation of Quantum Einstein Gravity in a higher-derivative truncation. Phys. Rev. D **66**(2), 025026 (2002). https://doi.org/10.1103/PhysRevD.66.025026. eprint: hep-th/0205062 (cit. on p. 74)

53. K. Groh, F. Saueressig, Ghost wavefunction renormalization in asymptotically safe quantum gravity. J. Phys. A Math. Gen. **43**, 365403 (2010). https://doi.org/10.1088/1751-8113/43/36/ 365403. arXiv:1001.5032 [hep-th] (cit. on p. 74)

54. E. Manrique, M. Reuter, Bimetric truncations for Quantum Einstein Gravity and asymptotic safety. Ann. Phys. **325**, 785–815 (2010). https://doi.org/10.1016/j.aop.2009.11.009. arXiv:0907.2617 [gr-qc] (cit. on p. 74)

55. E. Manrique, M. Reuter, F. Saueressig, Matter induced bimetric actions for gravity. Ann. Phys. **326**, 440–462 (2011). https://doi.org/10.1016/j.aop.2010.11.003. arXiv:1003.5129 [hep-th] (cit. on p. 74)

56. D. Becker, M. Reuter, En route to background independence: broken split-symmetry, and how to restore it with bi-metric average actions. Ann. Phys. **350**, 225–301 (2014). https://doi.org/ 10.1016/j.aop.2014.07.023. arXiv:1404.4537 [hep-th] (cit. on pp. 74, 86)

57. H. Gies, B. Knorr, S. Lippoldt, Generalized parametrization dependence in quantum gravity. Phys. Rev. D **92**(8), 084020 (2015). https://doi.org/10.1103/PhysRevD.92.084020. arXiv:1507.08859 [hep-th] (cit. on pp. 74, 86)

58. A. Eichhorn, H. Gies, M.M. Scherer, Asymptotically free scalar curvature ghost coupling in Quantum Einstein Gravity. Phys. Rev. D **80**(10), 104003 (2009). https://doi.org/10.1103/ PhysRevD.80.104003. arXiv:0907.1828 [hep-th] (cit. on p. 74)

59. S. Rechenberger, F. Saueressig, R2 phase diagram of Quantum Einstein Gravity and its spectral dimension. Phys. Rev. D **86**(2), 024018 (2012). https://doi.org/10.1103/PhysRevD.86.024018. arXiv:1206.0657 [hep-th] (cit. on p. 74)

60. A. Eichhorn, H. Gies, Ghost anomalous dimension in asymptotically safe quantum gravity. Phys. Rev. D **81**(10), 104010 (2010). https://doi.org/10.1103/PhysRevD.81.104010. arXiv:1001.5033 [hep-th] (cit. on p. 74)

61. A. Nink, M. Reuter, On the physical mechanism underlying asymptotic safety. J. High Energy Phys. **1**, 62 (2013). https://doi.org/10.1007/JHEP01(2013)062. arXiv:1208.0031 [hep-th] (cit. on p. 74)

62. D. Becker, M. Reuter, Towards a C-function in 4D quantum gravity. J. High Energy Phys. **3**, 65 (2015). https://doi.org/10.1007/JHEP03(2015)065. arXiv:1412.0468 [hep-th] (cit. on p. 74)

63. D. Becker, M. Reuter, Propagating gravitons vs. 'dark matter' in asymptotically safe quantum gravity. J. High Energy Phys. **12**, 25 (2014). https://doi.org/10.1007/JHEP12(2014)025. arXiv:1407.5848 [hep-th] (cit. on p. 74)

64. T. Aida, Y. Kitazawa, Two-loop prediction for scaling exponents in $2 + \epsilon$-dimensional quantum gravity. Nucl. Phys. B **491**, 427–458 (1997). https://doi.org/10.1016/S0550-3213(97)00091-6. eprint: hep-th/9609077 (cit. on pp. 84, 85)

65. H.W. Hamber, R.M. Williams, Simplicial quantum gravity in three dimensions: analytical and numerical results. Phys. Rev. D **47**, 510–532 (1993). https://doi.org/10.1103/PhysRevD.47.510 (cit. on p. 85)

66. H.W. Hamber, R.M. Williams, Discrete Wheeler-DeWitt equation. Phys. Rev. D **84**(10), 104033 (2011). https://doi.org/10.1103/PhysRevD.84.104033. arXiv:1109.2530 [hep-th] (cit. on p. 85)

67. H.W. Hamber, R. Toriumi, R.M. Williams, Wheeler-DeWitt equation in $2 + 1$ dimensions. Phys. Rev. D **86**(8), 084010 (2012). https://doi.org/10.1103/PhysRevD.86.084010. arXiv:1207.3759 [hep-th] (cit. on p. 85)

68. M. Reuter, H. Weyer, Quantum gravity at astrophysical distances? in *JCAP 12*, 001 (2004), p. 001. https://doi.org/10.1088/1475-7516/2004/12/001. eprint: hep-th/0410119 (cit. on p. 87)

Part III
Astrophysical and Cosmological Implications of Asymptotic Safety

Chapter 5
Inflationary Cosmology from Quantum Gravity-Matter Systems

The Asymptotic Safety Theory for Quantum Gravity provides a natural framework for the description of gravity within the context of Quantum Field Theory. As the resulting effective action smoothly interpolates between the short and long-distance regimes, the FRG also provides an ideal tool to study phenomenological implications of Quantum Gravity. The short distances modifications of gravity may in fact have several implications in astrophysics and cosmology. For instance, it is possible that quantum gravitational effects may be detected in the measurements of the Cosmic Microwave Background (CMB) radiation and in observations of the large scale structure of the Universe [1].

In the simple case of the Einstein–Hilbert truncation, several investigations have focused on the implications of the running of the Newton's constant in models of the early universe. In particular, the scale-dependent couplings can be encoded in a self-consistent manner by renormalization group improving the Einstein equations [2–6]. In particular, it has been shown that the renormalization group induced evolution of the Newton's constant and cosmological constant can provide a consistent cosmic history of the universe from the initial singularity to the current phase of accelerated expansion [2, 3, 5, 7, 8]. Moreover, the scaling properties of the 2-points correlation function of the graviton near the NGFP induce a scale invariant spectrum of the primordial perturbations, characterized by a spectral index n_s which, to a very good approximation, must satisfy $n_s \sim 1$ [5]. An effective Lagrangian encoding the leading quantum gravitational effects near the NGFP has been proposed in [9] within the Einstein–Hilbert truncation and a family of inflationary solutions has been found.

This chapter is based on the following publications:
- A. Bonanno, A. Platania-*Asymptotically Safe inflation from quadratic gravity*-Phys. Lett. B 750 (2015) 638 [arXiv:1507.03375]
- A. Bonanno, A.Platania-*Asymptotically Safe $R + R^2$ gravity*-PoS(CORFU2015)159
- A. Bonanno, A.Platania, F. Saueressig-*Cosmological bounds on the field content of asymptotically safe gravity-matter models*. Phys. Lett. B (2018). https://doi.org/10.1016/j.physletb.2018.06.047.

© Springer Nature Switzerland AG 2018
A. B. Platania, *Asymptotically Safe Gravity*, Springer Theses,
https://doi.org/10.1007/978-3-319-98794-1_5

Following the strategy advocated in [9], in this chapter we will construct two classes of modified gravity models and we will report a systematic comparison of our results with the recent Planck data [10]. Firstly, we will discuss a family of inflationary models derived from quantum-gravity modifications of the well-known $(R + R^2)$ Starobinsky model [11] and it will be shown that the predictions of Asymptotic Safety are in agreement with the recent Planck data on CMB anisotropies. Secondly, a similar analysis will be carried out for the case of Quantum Gravity minimally coupled to matter fields. In particular we will see that combining the results obtained in Sect. 4.3.2 with the recent Planck data can put constraints on the primordial matter content of the universe.

5.1 CMB Radiation and Cosmic Inflation

The discovery of the CMB, namely the radiation emitted by the last scattering surface, represents nowadays the most important evidence of the validity of the Cosmological Principle. Its spectrum reproduces a perfect black body radiation at a temperature $T \sim 2.7$ K, which is (almost) uniform in all directions. As the CMB provides a redshifted picture of the universe at the decoupling era, the primordial universe should have been extraordinary isotropic and homogeneous. Moreover, measurements of the density parameter suggest that our universe is nearly spatially flat. All these peculiar features reflect in a very special set of initial conditions at the decoupling era, and result in a fine-tuning problem. The inflationary scenario provides a possible solution to this problem. The latter is based on a primordial phase of accelerated expansion, by means of which the original quantum fluctuations have been smoothed out, resulting in small density fluctuations at the time of decoupling (see [12] for a review). According to this scenario, the density fluctuations at the last scattering surface served as "seeds" for the formation of all large-scale structures in the universe. Moreover, the occurrence of these primordial irregularities indirectly provides an explanation for the tiny anisotropies, $\delta T / T \sim 10^{-5}$, detected in the CMB radiation.

The power spectrum of CMB anisotropies provides information on the density fluctuations at the last scattering surface, and it is intrinsically related to the physics of the very early universe. The main features of the CMB power spectrum are encoded in the spectral index n_s and tensor-to-scalar ratio r. Denoting the power spectrum of scalar and tensorial perturbations by $\mathcal{P}_\zeta(k)$ and $\mathcal{P}_h(k)$ respectively, where k is a Fourier mode, and approximating their scaling by a power law with constant exponents

$$\mathcal{P}_\zeta(k) \simeq \mathcal{A}_s \left(\frac{k}{k_*} \right)^{n_s - 1} , \qquad \mathcal{P}_h(k) \simeq \mathcal{A}_t \left(\frac{k}{k_*} \right)^{n_t} , \qquad (5.1)$$

the spectral index and tensor-to-scalar ratio read

$$n_s = 1 + \frac{d(\log \mathcal{P}_\zeta)}{d(\log k)} \Bigg|_{k_*} , \qquad r = \frac{\mathcal{P}_h}{\mathcal{P}_\zeta} \Bigg|_{k_*} . \qquad (5.2)$$

Here $k_* \sim 0.05\,\mathrm{Mpc}^{-1}$ is a *pivot scale*, while $\mathscr{A}_s$ and $\mathscr{A}_t$ represent the power spectra amplitudes (see [12, 13] for details). The recent Planck data suggest that the power spectrum of scalar perturbations is *nearly* scale invariant, $n_s \sim 1$, but the exact equality $n_s = 1$ seems to be excluded [10]. As we will see in Sect. 5.1.1, the spectral index and tensor-to-scalar ratio can be theoretically evaluated by means of the slow-roll approximation and the comparison with the values provided by the Planck collaboration constitutes a constistency test for any cosmological model (see [13] for an overview of the most important inflationary models and their predictions).

5.1.1 Cosmic Inflation and Slow-Roll Approximation

Cosmic inflation can be described by means of a scalar field ϕ (inflaton) minimally coupled to gravity and subject to a potential $V(\phi)$. In particular, the scalar field is assumed to depend upon the cosmic time only and its potential can be thought as an effective vacuum energy, whose negative pressure may provide a period of accelerated expansion. The dynamics of the scalar field ϕ is determined by the analytical properties of $V(\phi)$ and is described by the Friedmann and conservation equations

$$H^2 = \frac{\kappa}{3}\left\{\dot{\phi}^2 + V(\phi)\right\}, \qquad \ddot{\phi} + 3H\dot{\phi} + V'(\phi) = 0 \qquad (5.3)$$

where $H(t) = \dot{a}/a$ stands for the Hubble parameter, $a(t)$ denotes the scale factor and $\kappa \equiv 8\pi G_N$, G_N being the observed value of the Newton's constant. In this description, the inflaton field gives rise to a phase of accelerated expansion when its kinetic energy is negligible respect to its potential (the field "slow-rolls"), namely $\dot{\phi}^2 \ll 2\,V(\phi)$ and $\ddot{\phi} \ll 3H\dot{\phi} + V'(\phi)$. Defining the slow-roll parameters as

$$\varepsilon(\phi) \equiv \frac{1}{2\kappa}\left(\frac{V'(\phi)}{V(\phi)}\right)^2, \qquad \eta(\phi) \equiv \frac{1}{\kappa}\left(\frac{V''(\phi)}{V(\phi)}\right), \qquad (5.4)$$

the quasi–de Sitter behavior is realized if $\varepsilon \ll 1$ and $\eta \ll 1$. Under this approximation inflation is assumed to end when $\varepsilon(\phi_f) = 1$ (or $\eta(\phi_f) = 1$), with $\phi_f = \phi(t_f)$. The initial state $\phi_i = \phi(t_i)$ is implicitly defined by the number of e-folds $N(\phi_i) \equiv \log[a(t_f)/a(t_i)]$

$$N(\phi_i) = \int_{t_i}^{t_f} H(t)\,dt = \kappa \int_{\phi_f}^{\phi_i} \frac{V(\phi)}{V'(\phi)}\,d\phi \qquad (5.5)$$

characterizing the period of exponential growth. On cosmological grounds one requires $N \gtrsim 55$ [12]. The slow-roll approximation can be used to quantitatively characterize the power spectrum of CMB anisotropies. In particular, the amplitude of the primordial scalar power spectrum can be written as [13]

$$\mathscr{A}_s = \frac{\kappa^2\,V(\phi_i)}{24\pi^2\,\varepsilon(\phi_i)} \qquad (5.6)$$

while the spectral index n_s and the tensor-to-scalar ratio r read [12, 13]

$$n_s = 1 - 6\,\varepsilon(\phi_i) + 2\,\eta(\phi_i) \ ,$$

$$r = 16\,\varepsilon(\phi_i) \ . \tag{5.7}$$

These expressions allow to easily compare the theoretical predictions of single-field inflationary models with the corresponding results provided by the Planck collaboration.

At this stage it is worth pointing out that the scalar degree of freedom associated with the inflaton field ϕ, can also be obtained by means of a $f(R)$ theory (see [14, 15] for an extensive review). In fact, a generic $f(R)$ theory can be though as a scalar-tensor theory described by the action

$$S = \frac{1}{16\pi\,G_N} \int \sqrt{-g}\,\left[f'(\chi)(R - \chi) + f(\chi) \right] d^4x \ . \tag{5.8}$$

The auxiliary scalar field χ satisfies the equation of motion $f''(\chi)(R - \chi) = 0$ and hence, provided that $f''(\chi) \neq 0$, the underlying $f(R)$ theory is equivalent to the scalar-tensor theory (5.8). In particular, the above action can be put in canonical form (i.e. with a standard propagating scalar field) by means of the conformal transformation

$$g_{\mu\nu} \longrightarrow g^E_{\mu\nu} = \varphi\, g_{\mu\nu} \ , \qquad \varphi \equiv f'(\chi) = e^{\sqrt{2\kappa/3}\,\phi} \ . \tag{5.9}$$

Here $\varphi \equiv \Omega^2$ is the conformal factor inducing the transformation of the theory (5.8) from the Jordan to the Einstein frame and $g^E_{\mu\nu}$ denotes the spacetime metric in the latter frame. More generally, all physical quantities in the Einstein frame will be labeled by a subscript "E". Applying the transformation law (5.9) to the action (5.8) finally yields

$$S = \int \sqrt{-g_E}\,\left[\tfrac{1}{2\kappa} R_E + \tfrac{1}{2} g^{\mu\nu}_E\, \partial_\mu\phi\, \partial_\nu\phi - V(\phi) \right] d^4x \tag{5.10}$$

where R_E denotes the scalar curvature in the Einstein frame, and the scalar potential $V(\phi)$ is given by the following expression

$$V[\varphi(\phi)] = \tfrac{1}{2\kappa} \left(\varphi\, \chi(\varphi) - f[\chi(\varphi)] \right) \varphi^{-2} \ . \tag{5.11}$$

The scalar field ϕ is the extra degree of freedom associated with the original $f(R)$ action. Therefore, $f(R)$ theories may provide a suitable period of cosmic inflation, capable of solving the basic issues for which inflation has been introduced. The simplest and most successful model based on a modification of the Einstein–Hilbert action is the $(R + R^2)$–Starobinsky model [11]. The aim of the next section is to study the inflationary scenario arising from the Quantum Gravity modifications of the Starobinsky model, due to the scaling of couplings around the NGFP.

5.2 Asymptotically Safe Inflation from $(R + R^2)$-Gravity

In this section we discuss a class of inflationary models arising from Quantum Gravity modifications of the well-known Starobinsky model [11]. This study generalizes the analysis of [9] by the inclusion of the additional relevant direction associated to the R^2 operator. As a main result, it will be shown that the resulting RG-improved model, obtained by taking into account the scaling of the couplings around the NGFP, is in agreement with the recent Planck data on the power spectrum of temperature fluctuations in the CMB radiation.

5.2.1 Motivation and Strategy

According to standard cosmology, the primordial quantum fluctuations were exponentially stretched during inflation. This resulted in small density perturbations at the decoupling era. In the context of Asymptotically Safe Gravity these quantum fluctuations can be identified with the fluctuations of the spacetime geometry occurring during the NGFP regime [5]. The anisotropies in the CMB, due to the density fluctuations at the last scattering surface, can thus be traced back to the Quantum Gravity effects occurring during the Planck era. In this scenario it is important to understand if QEG can produce a suitable inflationary scenario compatible with the observational data. In [5] it has been argued that QEG, in the Einstein–Hilbert approximation, naturally provides a period of power-law inflation caused by the running of the cosmological constant in the NGFP regime. Subsequently, when the RG evolution exits the NGFP regime, the energy density associated to the cosmological constant drops below that of ordinary matter and inflation ends. This scenario predicts an almost scale-invariant power spectrum $n_s \simeq 1$ without requiring an inflaton field.

Although the Einstein–Hilbert action provides a successful description of gravity at a classical level (low-energy regime), the trans-Planckian regime may also include higher-derivative contributions. In particular, if the gravitational Lagrangian has a $f(R)$ structure, the leading contribution at the NGFP should be $\mathcal{L}_{\text{NGFP}} \sim R^2$ [16]. Therefore the relevant direction associated to the R^2 operator may play a fundamental role in the description of the early universe cosmology. The aim of this section is to study the quantum corrections to the Starobinsky model arising from QEG and compare the corresponding inflationary models with the Planck data [10].

In order to include the quantum effects arising from QEG into the classical Starobinsky model we will employ a standard strategy used in Quantum Field Theory, i.e. the Renormalization Group (RG) improvement. According to this method, the leading quantum corrections can be included by replacing the classical coupling constants $\{g\}$ appearing in the action by the running couplings $\{g_k\}$ obtained as solutions of the RG equations. The resulting model provides an effective description of the system at the momentum scale k. At last, in order to close the system, one has to identify this energy scale with a characteristic dynamical scale of the system. This "cutoff identification" is an essential step to link the RG evolution with the dynamical evolution of the system under consideration. As an illustrative and important example, this method has been used to compute radiative corrections in scalar electrodynamics: the Coleman-Weinberg effective potential [17] can in fact be obtained by RG-improving the classical potential and identifying the RG scale k with the scalar field itself. In the case of gauge theories this scale identification must be generalized in a way that preserves gauge invariance. Therefore the infrared cutoff must be constructed from gauge invariant quantities. Accordingly, one possibility is to RG-improve the classical Lagrangian and identify the cutoff with the field strength $(F_{\mu\nu}F^{\mu\nu})^{1/4}$. As pointed out in [18], in the case of Yang–Mills theories this strategy suffices to capture the most important features of the quantum theory under consideration. This method has extensively been employed in both Quantum Electrodynamics and Quantum Chromodynamics [18–22].

Motivated by the successes at the level of Yang–Mills theory, a similar strategy can be employed to study the quantum gravity effects in actual gravitational phenomena. Over the years several works [4, 5, 23–27] have used the RG improvement procedures to study phenomenological implications of Asymptotic Safety. In the case of gravity the field strength is naturally provided by the spacetime curvature. One can therefore employ the cutoff identification $k^2 \sim R$. The RG-improved Lagrangian resulting from the latter scale setting is an effective $f(R)$-type Lagrangian whose general structure agrees very well with the high-curvature solution of the fixed point equation for a generic asymptotically safe $f(R)$ theory [9, 28–30].

5.2.2   Effective Inflationary Potential from Quadratic Gravity

Let us consider the following quadratic gravity Lagrangian

$$\mathcal{L}_k = \frac{k^2}{16\pi g_k}(R - 2\lambda_k k^2) - \beta_k R^2 \tag{5.12}$$

where g_k, λ_k and β_k are dimensionless running couplings whose infinite momentum limit is controlled by the NGFP in the $(g_k, \lambda_k, \beta_k)$ sub-theory space [31, 32]. The corresponding UV critical manifold is described by three relevant directions and the qualitative behavior of the renormalization group flow in the vicinity of the NGFP is rather simple. In particular there exist trajectories that emanate from the NGFP and possess a long classical regime where the effective action reduces to the standard Einstein–Hilbert action [32]. The Lagrangian defined in Eq. (5.12) classically matches the original Starobinsky model . The question we want to address here is whether the scale-dependence of the couplings induced by the NGFP affects the Starobinsky inflationary model [11] and if the RG-improved model is compatible with the Planck data on the power spectrum of CMB anisotropies.

Following the strategy described in Sect. 5.2.1, the first step to find a RG-improved effective action is to replace the running couplings $(g_k, \lambda_k, \beta_k)$ with the solutions of the FRGE. However, the flow equation associated with these running couplings cannot be solved analytically. Therefore, one can either use numerical methods or resort to some approximations. Since the study of inflationary scenario in the slow-roll approximation requires an analytical expression for the inflationary potential, we will adopt the latter strategy. First, we approximate the running of λ_k with its tree-level scaling $\lambda_k \sim c_2 k^{-2}$, where c_2 is a dimensionful constant. Although this approximation is quite drastic, it is justified by the impressive stability of the critical exponents for the Newton's constant against the inclusion of higher order truncations [33]. Moreover, it allows to decouple the running of g_k from the running of β_k, so that an analytical expression for the running Newton's constant [24] can be used. Secondly, we approximate the running coupling β_k with its scaling around the NGFP. Under these assumptions the renormalized flow is encoded in the following running couplings [24]

$$g_k = \frac{(c_1\,\mu^{-2})\,k^2}{1 + \omega\,(k^2 - \mu^2)\,(c_1\,\mu^{-2})}\,, \tag{5.13a}$$

$$\beta_k = \beta_* + b_0 \left(\frac{k^2}{\mu^2}\right)^{-\frac{\theta_3}{2}}. \tag{5.13b}$$

Here μ is an arbitrary scale defining $c_1 \equiv g(k = \mu)$, θ_3 is the critical exponent associated with the R^2 relevant direction, $\omega \equiv 1/g_*$ and the parameters $g_* = 6\pi/23$ and $\beta_* \sim 0.002$ [31] define the location of the NGFP in the subspace under consideration. It is important to stress that, as long as $c_1 < 6\pi/23$, the running described by (5.13a) smoothly interpolates between the GFP and the NGFP, and therefore it captures the qualitative features of the flow described in [32]. Finally, the constants b_0, c_1 and c_2 are free parameters, corresponding to the three relevant directions of the UV critical surface, and must be fixed by confronting the model with observations. In other words, by changing b_0, c_1 and c_2 it is possible to explore various RG trajectories all ending at the NGFP. The relevant question is if it is possible to actually constrain these numbers, in particular the value of b_0.

As it will be important for our discussions, we note here that Eq. (5.13a) can be used to write down the expression for the dimensionful running Newton's coupling $G_k = k^{-2} g_k$. Recalling that $c_1 \equiv g(k = \mu)$, one can identify $G_\mu \equiv c_1 \mu^{-2}$. Rewriting (5.13a) in terms of G_μ and fixing the renormalization scale to $\mu = 0$, the running Newton's constant is compactly written as [24]

$$G_k = \frac{G_0}{1 + \omega\,G_0\,k^2} \tag{5.14}$$

where G_0 is the infrared value of the Newton's constant. Since in the Einstein–Hilbert truncation G_k is constant during the entire classical regime, including the observational scale $k_{\mathrm{obs}} > 0$, in this approximation G_0 coincides with the observed Newton's constant $G_N \equiv G(k_{\mathrm{obs}})$. On the other hand, if a Quasi Fixed Point mechanism governs the infrared limit, G_k may undergo a further running and its value would deviate from the observed one, as we discussed in Sect. 4.4. Such a behavior is not included in (5.14). Finally it is important to remark that the dimensionful Newton's coupling (5.14) vanishes for $k \to \infty$.

According to the arguments of Sect. 5.2.1, an RG-improved $f(R)$ effective action can be derived by substituting the running couplings (5.13) into Eq. (5.12) and by employing the cutoff identification [9, 29, 30]

$$k^2 \equiv \xi \, R \tag{5.15}$$

where ξ is an arbitrary positive constant. We thus obtain the following effective action

$$S_{\text{eff}} = \frac{1}{2\kappa} \int d^4x \sqrt{-g} \left[R + \tilde{\alpha} \, R^{2-\frac{\theta_3}{2}} + \frac{R^2}{6m^2} - 2\tilde{\Lambda} \right] . \tag{5.16}$$

Here $\kappa \equiv 8\pi \, G_N$, m^2, $\tilde{\alpha}$ and $\tilde{\Lambda}$ are effective coupling constants given by

$$\kappa = \frac{48 \, \pi^2 \, c_1}{6 \, \pi \, \mu^2 - 23 \, (\mu^2 + 2\,\xi \, c_2) \, c_1} \, , \qquad m^2 = \frac{8\pi^2}{\kappa \, (23\,\xi - 96\pi^2 \beta_*)} \, , \tag{5.17a}$$

$$\tilde{\Lambda} = \frac{\mu^2 (6\pi - 23 \, c_1) \, c_2}{6 \, \pi \, \mu^2 - 23 \, (\mu^2 + 2\,\xi \, c_2) \, c_1} \, , \qquad \tilde{\alpha} = -2 \, b_0 \, \kappa \, \left(\xi \mu^{-2} \right)^{-\frac{\theta_3}{2}} . \tag{5.17b}$$

In particular, the effective parameters κ and m^2 are positive definite if $\xi > 96\pi^2 \beta_*/23$ and $c_2 < \mu^2 (6\pi - 23c_1)/46\xi c_1$. Notably, the coupling constant α only depends on θ_3 and b_0. In particular the numerical evidence accumulated so far has shown that the value of θ_3 is rather stable against the introduction of higher order derivatives in the flow equation [34], as it is expected for a critical exponent. On the contrary b_0 is by construction a non-universal quantity whose value cannot be determined a priori. It labels a specific trajectory emanating from the fixed point and its actual value should be determined by matching with a low-energy observable.

In order to obtain the inflationary potential (5.11) associated with the RG-improved action (5.16), we introduce the auxiliary field φ defined via

$$\varphi(\chi) \equiv 1 + \tilde{\alpha} \left(2 - \frac{\theta_3}{2} \right) \chi^{1-\frac{\theta_3}{2}} + \frac{\chi}{3m^2} . \tag{5.18}$$

Due to its non-linearity, the task of inverting (5.18) can be very difficult and one should resort to numerical methods. However, according to the analysis of [33], the critical exponent θ_3 is rather close to unity. Hence, in order to obtain analytic expressions, we can approximate $\theta_3 = 1$ for any practical calculation. In this case we explicitly obtain the two branches

$$\chi_\pm = \frac{3}{8} \left(27\tilde{\alpha}^2 m^4 + 8m^2(\varphi - 1) \pm 3\sqrt{3} \sqrt{27\tilde{\alpha}^4 m^8 + 16\tilde{\alpha}^2 m^6(\varphi - 1)} \right) \tag{5.19}$$

with the reality condition $\varphi \geq 1 - 27m^2\tilde{\alpha}^2/16$. Using these solutions we can obtain a canonically coupled scalar field by introducing the conformal metric (5.9).[1] In this way the effective action (5.16), which is of the $f(R)$–type in the Jordan frame, looks like a scalar-tensor theory in the Einstein frame and finally reads

$$S_{\text{eff}} = \int d^4x \sqrt{-g_E} \left[\frac{1}{2\kappa} R_E - \frac{1}{2} g_E^{\mu\nu} \partial_\mu \phi \partial_\nu \phi - V_\pm(\phi) \right] \qquad (5.20)$$

where the effective scalar potential $V_\pm(\phi)$ is given by

$$
\begin{aligned}
V_\pm(\phi) = \frac{m^2 e^{-2\sqrt{\frac{2\kappa}{3}}\phi}}{256\,\kappa} &\left\{ 192 \left(e^{\sqrt{\frac{2\kappa}{3}}\phi} - 1 \right)^2 - 3\alpha^4 + 128\Lambda \right. \\
&- \sqrt{32}\alpha \left[\left(\alpha^2 + 8e^{\sqrt{\frac{2\kappa}{3}}\phi} - 8 \right) \pm \alpha\sqrt{\alpha^2 + 16 e^{\sqrt{\frac{2\kappa}{3}}\phi} - 16} \right]^{\frac{3}{2}} \\
&\left. - 3\alpha^2 \left(\alpha^2 + 16 e^{\sqrt{\frac{2\kappa}{3}}\phi} - 16 \right) \mp 6\alpha^3 \sqrt{\alpha^2 + 16 e^{\sqrt{\frac{2\kappa}{3}}\phi} - 16} \right\}
\end{aligned} \qquad (5.21)
$$

and we have measured $\tilde{\alpha}$ and $\tilde{\Lambda}$ in units of the scalaron mass m by introducing the dimensionless quantities $\Lambda = m^{-2} \tilde{\Lambda}$ and $\alpha = 3\sqrt{3}m\,\tilde{\alpha}$.

5.2.3 Inflaton Dynamics and Primordial Power Spectrum

With the aim of discussing inflation in the slow-roll approximation, it is important to study the analytical properties of the family of potentials $V_\pm(\phi)$ (see Figs. 5.1 and 5.2). We firstly notice that for all values (α, Λ) the scalar potentials $V_\pm(\phi)$ in Eq. (5.21) are characterized by a plateau

$$\lim_{\phi \to +\infty} V_\pm(\phi) = \frac{3m^2}{4\kappa} . \qquad (5.22)$$

In order to verify the slow-roll conditions, the inflaton field must start its motion from $\phi > M_{\text{pl}}$ in a quasi–de Sitter state and then proceed towards $\phi \ll M_{\text{pl}}$. The shape of the potential for $\phi \ll M_{\text{pl}}$, and thus the inflation dynamics, strongly depends on the values (α, Λ). In our case only two behaviors are possible

- $V_\pm(\phi)$ develops a minimum for $\phi \ll M_{\text{pl}}$ and $\lim_{\phi \to -\infty} V_\pm(\phi) = +\infty$
- $V_\pm(\phi)$ has no stationary points and $\lim_{\phi \to -\infty} V_\pm(\phi) = -\infty$

---

[1] Note that the field redefinition employed in going from the Jordan to the Einstein frame involves new additional contributions in the path-integral arising from the Jacobians. In investigations including the presence of matter field in the starting Lagrangian, these new terms cannot be neglected.

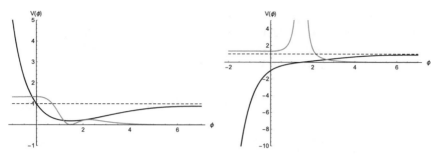

Fig. 5.1 Inflationary potential $V_-(\phi)$ (black line) and slow-roll function $\varepsilon(\phi)$ (gray line) for $\alpha = -10$ and $\Lambda = 2$ (left panel) and $\Lambda = -2$ (right panel). The dashed line corresponds to the limiting condition $\varepsilon(\phi) = 1$, which separates the slow-roll regime ($\varepsilon(\phi) < 1$) from non-inflationary phases. Potentials with $V_\pm(\phi_{min}) > 0$ violate the slow-roll condition $\varepsilon(\phi) < 1$ at most for a negligible time period, thus giving rise to eternal inflation. On the other hand, potential without a minimum can have a "graceful exit" from inflation, but no standard reheating phase is possible

In particular, if $V_\pm(\phi)$ has a minimum then its value $V_\pm(\phi_{min})$ is always positive for $V_-(\phi)$, while it can be either positive or negative for $V_+(\phi)$.

The post-inflationary reheating phase due to damped oscillations of the inflaton is clearly possible only when $V_\pm(\phi)$ has a minimum. In our case inflation can naturally end by violation of the slow-roll condition $\varepsilon(\phi_f) = 1$ only when the minimum value of the potential is negative, $V(\phi_{min}) \leq 0$. In fact, when $V_\pm(\phi_{min}) > 0$ the slow-roll condition $\varepsilon = 1$ is verified at $\phi_f < \phi_{min}$; subsequently, when the inflaton starts oscillating around ϕ_{min}, the field ϕ enters back the slow-roll regime $\varepsilon(\phi) < 1$ and remains in this region. Hence, this class of potentials (Fig. 5.1, left panel) gives rise to *eternal inflation*. On the contrary, the family of potentials which are unbounded from below is characterized by a well defined exit from inflation (Fig. 5.1, right panel), but the reheating phase cannot be explained by inflaton parametric oscillations.

In what follows we will restrict our attention to the class of potentials providing a well defined exit from inflation, followed by a phase of parametric oscillations of the field ϕ. These particular features are realized by the family of potentials $V_+(\phi)$ for $\alpha \in [1, 3]$ and $\Lambda \in [0, 1.5]$. This case is illustrated in Fig. 5.2 for various values of α and $\Lambda = 1.4$. Notably, although for α and Λ very close to zero the potential (5.21) is only a small modification of the classical Starobinsky model, for $\alpha \in [1, 3]$ and $\Lambda \in [0, 1.5]$ the potential $V_+(\phi)$ develops a non-trivial minimum at negative values of the potential which makes the quantum-corrected $(R + R^2)$–Starobinsky model significantly different from the original one.

At this point, it is important to compare the main features of the primordial power spectrum arising from the RG-improved inflationary potential (5.21) with the data provided by the Planck collaboration. These important properties are encoded in the spectral index n_s and tensor-to-scalar ratio r, Eq. (5.7). In particular, the evaluation of such parameters requires the knowledge of the initial value of the inflaton field, $\phi_i = \phi(t_i)$, which can be obtained by inverting (5.5) once the number of e-folds N is fixed. However, while the final value of the field $\phi_f = \phi(t_f)$ can be analytically determined by solving $\varepsilon(\phi) = 1$ (violation of the slow-roll condition), in our case

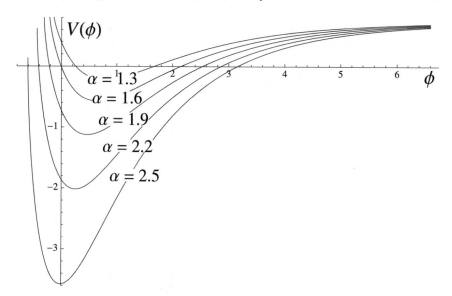

Fig. 5.2 Inflationary potential $V_+(\phi)$ for various values α and $\Lambda = 1.4$. In this case inflation lasts for a finite amount of time and a phase of parametric oscillations of the field ϕ follows

Fig. 5.3 We compare the theoretical predictions of our model with the Planck collaboration 2015 data release assuming ΛCDM [10]. The different theoretical "points" in the (r, n_s)–plane are obtained by fixing $\Lambda = 1.4$ and varying the value of α. In particular, triangles are for $N = 55$ and squares for $N = 60$ e-folds. Solid and dashed lines are the 1σ and 2σ confidence levels, respectively

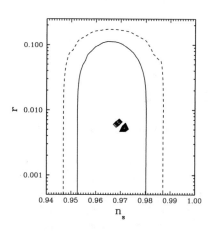

the evaluation of the initial value ϕ_i requires numerical methods. The values of the spectral index n_s and tensor-to-scalar ratio r resulting from the class of potentials under consideration are displayed in Table 5.1. Confronting n_s and r with the Planck data [10], the present RG-improved model agrees very well, as it is shown in Fig. 5.3. Notably, the present model significantly differs from the well known Starobinsky model [11], as it predicts a tensor-to-scalar ratio which is significantly higher. Finally, according to Eq. (5.6), the normalization of the scalar power spectrum at the pivot scale $k_* = 0.05\,\mathrm{Mpc}^{-1}$ provides us with $m \sim (1.5 - 7) \cdot 10^{14}\,\mathrm{GeV}$, depending on the value of α.

Table 5.1 Values of the spectral index n_s and tensor-to-scalar ratio r obtained from the RG-improved Starobinsky inflation for different values of (α, Λ) and number of e-folds N. In particular, the range of values obtained for the spectral index, $n_s \in [0.965, 0.972]$, is in agreement with the one obtained by the Planck Collaboration, $n_s = 0.968 \pm 0.006$, and the tensor-to-scalar ratio is always compatible with their upper limit, $r < 0.11$

Cases		$N = 50$		$N = 55$		$N = 60$	
Λ	α	n_s	r	n_s	r	n_s	r
0	1.0	0.965	0.0069	0.968	0.0058	0.971	0.0050
	1.8	0.966	0.0074	0.969	0.0063	0.972	0.0055
	2.6	0.967	0.0076	0.969	0.0065	0.972	0.0056
1	1.0	0.965	0.0070	0.968	0.0059	0.971	0.0051
	1.8	0.966	0.0074	0.969	0.0063	0.972	0.0055
	2.6	0.967	0.0076	0.969	0.0065	0.972	0.0056

Table 5.2 Values of the spectral index n_s and tensor-to-scalar ratio r obtained from the class of potentials (5.21) without stationary points. Although the latter cannot provide a standard reheating phase, the values of n_s and r are compatible with the recent Planck data

Cases		$N = 50$		$N = 55$		$N = 60$	
Λ	α	n_s	r	n_s	r	n_s	r
-5	-2.0	0.807	$1.5 \cdot 10^{-6}$	0.808	$6.0 \cdot 10^{-7}$	0.808	$2.3 \cdot 10^{-7}$
	$+1.0$	0.966	0.0066	0.969	0.0056	0.972	0.0048
	$+2.0$	0.966	0.0074	0.969	0.0064	0.972	0.0055
-10	-2.0	0.809	$1.6 \cdot 10^{-6}$	0.810	$6.1 \cdot 10^{-7}$	0.810	$2.4 \cdot 10^{-7}$
	$+1.0$	0.967	0.0063	0.970	0.0054	0.972	0.0047
	$+2.0$	0.967	0.0074	0.970	0.0063	0.972	0.0054

As a final remark, the class of potentials unbounded from below cannot produce a standard reheating phase but the well defined exit from inflation allows to determine n_s and r within the usual slow-roll approximation. A representative set of values (n_s, r) derived from this class of potentials is reported in Table 5.2. Although the way reheating occurs in these cases is beyond the purpose of this thesis, it is instructive to compare the values (n_s, r) resulting from the families of potentials with and without a local minimum. Notably, when $\alpha > 0$ the class of potentials without minima gives rise to spectral indeces and tensor-to-scalar ratios similar to the ones reported in Table 5.1.

5.2.4 Post-inflationary Dynamics

After the end of inflation, the inflaton field ϕ starts oscillating about the minimum $\phi_{\min}$ of its potential $V_+(\phi)$. In order to study this phase, we can approximate

$$V_+(\phi) \sim V(\phi) = \frac{b}{2}\left[(\phi - \phi_{\min})^2 - c\right] \tag{5.23}$$

where $\phi_{\min}(\alpha, \Lambda)$, $b(\alpha, \Lambda) = V''_+(\phi_{\min})$ and $c(\alpha, \Lambda) = -2\, V_+(\phi_{\min})/V''_+(\phi_{\min})$
depend on both α and Λ, and are given by the following expressions

$$\phi_{\min}(\alpha, \Lambda) = \frac{\sqrt{\tfrac{3}{2}}\left(3\alpha^3\left(\alpha^2 - 4\right) - 32\alpha\Lambda + 4\left(\alpha^2 - 6\right)|\alpha|^3\right)}{6\alpha\left(\alpha^2 - 8\right)\left(\alpha^2 + 2\right) - 64\alpha\Lambda + 8\left(\alpha^2 - 9\right)|\alpha|^3},\tag{5.24}$$

$$b(\alpha, \Lambda) = \frac{48 + 18\alpha^2 - 3\alpha^4 + 32\Lambda - 4\alpha^3|\alpha| + 36\alpha|\alpha|}{24},\tag{5.25}$$

$$\begin{aligned}
c(\alpha, \Lambda) =\ & \frac{8\alpha\left(15\alpha^4 - 3\alpha^6 - 96\Lambda + 8\alpha^2(15 + 4\Lambda)\right)|\alpha|}{\tfrac{8}{3}\left(48 + 18\alpha^2 - 3\alpha^4 + 32\Lambda - 4\alpha\left(\alpha^2 - 9\right)|\alpha|\right)^2} \\[2mm]
& + \frac{-25\alpha^8 + 132\alpha^6 - 384\alpha^2\Lambda}{\tfrac{8}{3}\left(48 + 18\alpha^2 - 3\alpha^4 + 32\Lambda - 4\alpha\left(\alpha^2 - 9\right)|\alpha|\right)^2} \\[2mm]
& + \frac{48\alpha^4(21 + 4\Lambda) - 1024\Lambda(3 + \Lambda)}{\tfrac{8}{3}\left(48 + 18\alpha^2 - 3\alpha^4 + 32\Lambda - 4\alpha\left(\alpha^2 - 9\right)|\alpha|\right)^2}.
\end{aligned}\tag{5.26}$$

The time evolution of the field $\phi(t)$ is described by the conservation equation

$$\ddot{\phi} + 3\,H\,\dot{\phi} + V'(\phi) = 0\tag{5.27}$$

where the Hubble parameter $H(t)$ is given by

$$H^2 \equiv \frac{\kappa}{3}\left(\frac{1}{2}\dot{\phi}^2 + V(\phi)\right) = \frac{\kappa}{6}\left[\dot{\phi}(t)^2 + b\big(\phi(t) - \phi_{\min}\big)^2 - bc\right].\tag{5.28}$$

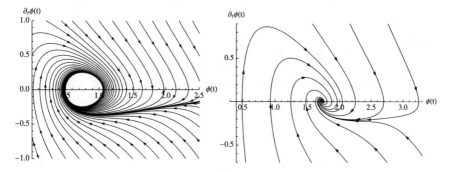

Fig. 5.4 Phase diagrams describing the dynamics of the inflaton field in the $(\phi, \dot{\phi})$–plane. The figure in the left panel is obtained for $\alpha = 1$ and $\Lambda = 1$ and corresponds to the case $V_+(\phi_{\min}) \leq 0$. The asymptotic behavior is controlled by a limit cycle. In contrast, the phase diagram in the right panel, obtained setting $\alpha = 1.5$ and $\Lambda = 10$, is characterized by an attractive fixed point $(\phi_{\min}, 0)$. This case arises when $V_+(\phi_{\min}) > 0$

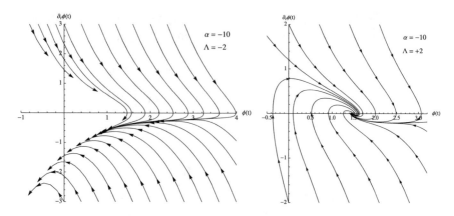

Fig. 5.5 Phase diagrams in the $(\phi, \dot{\phi})$–plane describing the inflationary dynamics for the class of potentials $V_-(\phi)$. The left panel describes the dynamics arising from unbounded potentials. The right panel corresponds to the case of potentials with a minimum and $V_-(\phi_{min}) > 0$. The corresponding dynamics matches the one illustrated in the right panel of Fig. 5.4

Putting $x(t) = \sqrt{b}\,(\phi(t) - \phi_{min})$ and $y(t) = \dot{\phi}(t)$, Eq. (5.27) can be written as

$$\begin{cases} \dot{y} = -\left[\frac{3\kappa}{2}\left(y^2 + x^2 - bc\right)\right]^{\frac{1}{2}} y - \sqrt{b}\, x \ , \\ \dot{x} = \sqrt{b}\, y \ . \end{cases} \tag{5.29}$$

The long time behavior of this dynamical system is mainly determined by the sign of $z \equiv bc = -2\,V(\phi_{min})$. If $z \leq 0$ $(V(\phi_{min}) \geq 0)$ then the origin of the (x, y)–plane, namely the minimum $(\phi_{min}, V(\phi_{min}))$, is an attractive node. On the contrary, if $z > 0$ $(V(\phi_{min}) < 0)$ a limit cycle defined by $y^2 + x^2 = z$ appears. The limiting case $z = 0$ thus represents a Hopf bifurcation point. While these findings concern the asymptotic behavior of the inflaton field, its dynamics can be studied by solving Eq. (5.29). The phase diagrams corresponding to the cases $V_+(\phi_{min}) < 0$ and $V_+(\phi_{min}) \geq 0$ are depicted in Fig. 5.4. For comparison we also report a sample of phase diagrams obtained from the potential $V_-(\phi)$ in Fig. 5.5.

Finally, it is interesting to study the behavior of the scale factor $a(t)$ during the reheating era. In particular, combining the system (5.29) with Eq. (5.28) and using the Krylov–Bogolyubov averaging method [35, 36], we obtain

$$a(t) = \begin{cases} \left[\sin\left(\sqrt{\frac{3}{8}\,|z|}\,t\right)\right]^{2/3} & z > 0 \\ t^{2/3} & z = 0 \\ \left[\sinh\left(\sqrt{\frac{3}{8}\,|z|}\,t\right)\right]^{2/3} & z < 0 \end{cases} \tag{5.30}$$

In the limiting case $z = 0$ $(V(\phi_{min}) = 0$, Starobinsky model), the scale factor $a(t)$ describes a matter-dominated epoch, while the solutions with $V(\phi_{min}) \neq 0$ are compatible with a matter-dominated era only at the beginning of the oscillatory phase. On the other hand, a consistent treatment of the reheating phase must include the contribution of the matter fields. This is beyond the scope of this dissertation.

5.3 Constraints on Early Universe Matter Fields from Foliated Quantum Gravity-Matter Systems

The behavior of gravity in the high-energy regime is controlled by the NGFP of the renormalization group flow. Assuming that the Asymptotic Safety scenario is correct, the precise values of the critical exponents may play a fundamental role in the description of the primordial evolution of the universe. As we have seen, in the case of pure gravity the critical exponents are complex and conjugate, making the flow spiraling out from the NGFP. However, the critical exponents associated with the gravitational couplings also depend on the matter content of the theory and, according to our findings of Chap. 4, they should be real and positive for a wide class of models, including the Standard Model and its modifications. Although we have assumed the matter fields to be minimally coupled to gravity and non-interacting, this approximation may be suitable for the description of the very early universe. Following the strategy introduced in Sect. 5.2.1 and using the results of Sect. 4.3.2, in this section we will see that comparing the RG-improved effective action resulting from the flow of gravity-matter systems with the Planck data, provides important constraints on the primordial matter content of the theory.

In the present discussion we restrict the gravitational sector to the Einstein–Hilbert truncation. The corresponding scale-dependent Lagrangian reads

$$\mathcal{L}_k = \frac{k^2}{16\pi g_k}(R - 2\lambda_k k^2) \tag{5.31}$$

where the running couplings g_k and λ_k are the solutions of the beta functions (4.49). For our purpose it suffices to study the linearized flow around the NGFP, where the scaling behavior for λ_k and g_k is determined by the critical exponents θ_i

$$\lambda_k = \lambda_* + c_1 e_{11} k^{-\theta_1} + c_2 e_{21} k^{-\theta_2} \quad , \tag{5.32a}$$

$$g_k = g_* + c_1 e_{12} k^{-\theta_1} + c_2 e_{22} k^{-\theta_2} \quad . \tag{5.32b}$$

Here $e_{ij} \equiv (\mathbf{e}_i)_j$ are the eigenvectors of the stability matrix $\mathcal{S}_{ij} \equiv \partial_j \beta_i|_*$, and $(-\theta_i)$ are the corresponding eigenvalues. The only free parameters, c_1 and c_2, allow to select a particular RG trajectory, and can be determined by comparing physical observables (e.g. physical couplings in the effective Lagrangian) with observations.

Following Sect. 5.2.1, an RG-improved Lagrangian can be obtained by substituting the running couplings (5.32) into Eq. (5.31). The resulting analytical expression strongly depends on the critical exponents θ_i. According to the discussion in Sect. 4.3.2, the physically interesting gravity-matter models possess positive and real stability coefficients (see Table 4.4). In order to obtain an analytical expression for the effective Lagrangian, we make the approximation $\theta_i \in \mathbb{N}$. Assuming $\theta_1 > \theta_2$, an asymptotic expansion of g_k^{-1}, with g_k given by (5.32b), yields

$$g_k^{-1} = g_*^{-2}(g_* - c_2\, e_{22}\, k^{-\theta_2}) + O(k^{-\theta_2 - 1}) \ . \tag{5.33}$$

We can thus approximate the coefficient $G_k^{-1} = k^2 g_k^{-1}$ in the Lagrangian (5.31) with its leading asymptotic behavior

$$\frac{1}{G_k} \simeq g_*^{-2}(g_* - c_2\, e_{22}\, k^{-\theta_2})\, k^2 \ . \tag{5.34}$$

Finally, using (5.34) and setting $k^2 = \xi R$ as before, yields the following RG-improved Lagrangian

$$\mathcal{L}_{\mathrm{eff}} = a_0\, R^2 + b_1\, R^{\frac{4-\theta_1-\theta_2}{2}} + b_2\, R^{\frac{4-\theta_1}{2}} + b_3\, R^{\frac{4-\theta_2}{2}} + b_4\, R^{2-\theta_2} \ . \tag{5.35}$$

The coefficients b_i depend on the initial data, parametrized by c_i, the set of parameters (λ_*, g_*, ξ), the eigenvalues θ_i and the eigenvector components e_{ij}. They are given by the following expressions

$$b_1 = \frac{c_1 c_2 (e_{11}\, e_{22})\, \xi^{\frac{4-\theta_1-\theta_2}{2}}}{8\pi g_*^2} \ , \qquad\qquad b_2 = \frac{c_1 (-e_{11})\, \xi^{\frac{4-\theta_1}{2}}}{8\pi g_*} \ , \tag{5.36}$$

$$b_3 = \frac{c_2 (2\, e_{22}\xi\lambda_* - 2\, e_{21}\, \xi g_* - e_{22})\, \xi^{\frac{2-\theta_2}{2}}}{16\pi g_*^2} \ , \qquad b_4 = \frac{c_2^2 (e_{21}\, e_{22})\, \xi^{2-\theta_2}}{8\pi g_*^2} \ . \tag{5.37}$$

The coefficient of R^2 only depends on the parameters (λ_*, g_*, ξ) and reads

$$a_0 = \frac{\xi\,(1 - 2\xi\lambda_*)}{16\pi g_*} \ . \tag{5.38}$$

Notably, the RG improvement of the Einstein–Hilbert action always produces a R^2-term, independently of the precise matter content of the theory. In other words it does not depend on the stability coefficients θ_i.

At this point we must require the RG-improved Lagrangian (5.35) to produce an inflationary scenario in agreement with the Planck data [10]. In particular, the values of the critical exponents θ_i should give rise to a Starobinsky-like Lagrangian, with at least one critical exponent $\theta_i \leq 4$. In fact, if all critical exponents satisfy the condition $\theta_i > 4$ then all b_i-terms are suppressed and the resulting Lagrangian $\mathcal{L} \sim R^2$ gives rise to a scale invariant scalar power spectrum, $n_s = 1$, which is excluded by the Planck data. Therefore, as $\theta_1 > \theta_2$ is assumed, the first constraint we find reads

$$\theta_2 \leq 4 \ . \tag{5.39}$$

According to Table 4.4, the only models compatible with the latter condition are the Standard Model, including its modifications, and the MSSM. In particular for all these models the critical exponents are

$$\theta_1 \simeq 4 \ , \qquad \theta_2 \simeq 2 \ , \tag{5.40}$$

while the eigenvectors $\mathbf{e}_i$ read

$$\mathbf{e}_1 \simeq \begin{pmatrix} -1 \\ 0 \end{pmatrix} \ , \qquad \mathbf{e}_2 \simeq \begin{pmatrix} +1/2 \\ -1/2 \end{pmatrix} \ . \tag{5.41}$$

The resulting effective Lagrangian $\mathcal{L}_{\text{eff}}$ (up to subleading corrections in $1/k^2$) thus reads

$$\mathcal{L}_{\text{eff}} = \frac{c_2 \left(1 - 2g_*\xi - 2\lambda_*\xi\right)}{32\pi g_*^2} R + \frac{\xi(1 - 2\lambda_*\xi)}{16\pi g_*} R^2 - \frac{c_2^2 - 4c_1 g_*}{32\pi g_*^2} \ . \tag{5.42}$$

Hence, when the gravitational renormalization group flow is combined with SM-type theories, the RG-improved Einstein–Hilbert Lagrangian results in a $(R + R^2)$ theory.

Finally, in order to recast (5.42) in the standard Starobinsky form, we require the coefficients of R and R^2 to define the observed Newton's constant G_N and scalaron mass m

$$G_N = \frac{2g_*^2}{c_2 \left(1 - 2g_*\xi - 2\lambda_*\xi\right)} \ , \tag{5.43a}$$

$$m^2 = \frac{g_*}{6\xi \left(1 - 2\lambda_*\xi\right) G_N} \ . \tag{5.43b}$$

Solving Eq. (5.43a) with respect to ξ and inserting the result into Eq. (5.43b), we obtain a second order equation for c_2 which admits real solutions if

$$\tilde{m}^2(g_* + \lambda_*)^2 \left(3\tilde{m}^2 - 4g_*\lambda_*\right) > 0 \ . \tag{5.44}$$

where we introduced the dimensionless quantity $\tilde{m}^2 = G_N m^2 \equiv m^2/M_{\text{pl}}^2$. The value of the mass m is fixed by the CMB normalization of the power spectrum, $\tilde{m}^2 \propto \mathcal{A}_s \sim 10^{-10}$, and for a Starobinsky-like model this implies $\tilde{m} \sim 10^{-5}$ (see [13] for details). As a consequence, we obtain the following constraint

$$g_*\lambda_* \lesssim 0 \ . \tag{5.45}$$

Since g_* must be positive, only negative values of λ_* are allowed to produce a physically acceptable inflationary model. According to Table 4.4, the gravity-matter models compatible with this second constraint are again the SM (and its minor extensions) and the MSSM. It follows that, under the Einstein–Hilbert approximation, only a SM-like theory can provide a suitable period of inflation compatible with the Planck data. However, it is worth pointing out that the results in Table 4.4 slightly change in the metric approach to Asymptotic Safety [37, 38] and, although the constraints (5.39) and (5.45) are still valid if the Einstein–Hilbert truncation is assumed, in the metric approach the matter models compatible with the Planck data may be different.

References

1. J.J. Halliwell, S.W. Hawking, Origin of structure in the Universe. Phys. Rev. D **31**, 1777–1791 (1985). https://doi.org/10.1103/PhysRevD.31.1777 (cit. on p. 99)
2. A. Bonanno, M. Reuter, Cosmology of the Planck era from a renormalization group for quantum gravity. Phys. Rev. D **65**(4), 043508 (2002). https://doi.org/10.1103/PhysRevD.65.043508. arXiv:hep-th/0106133 (cit. on p. 99)
3. M. Reuter, F. Saueressig, From big bang to asymptotic de Sitter: complete cosmologies in a quantum gravity framework. JCAP **9**, 12 (2005). https://doi.org/10.1088/1475-7516/2005/09/012. arXiv:hep-th/0507167 (cit. on p. 99)
4. A. Bonanno et al., The accelerated expansion of the universe as a crossover phenomenon. Class. Quantum Gravity **23**, 3103–3110 (2006). https://doi.org/10.1088/0264-9381/23/9/020. eprint: astro-ph/0507670 (cit. on pp. 99, 105)
5. A. Bonanno, M. Reuter, Entropy signature of the running cosmological constant. JCAP **8**, 24 (2007). https://doi.org/10.1088/1475-7516/2007/08/024. arXiv:0706.0174 [hep-th] (cit. on pp. 99, 100, 104, 105)
6. G. D'Odorico, F. Saueressig, Quantum phase transitions in the Belinsky-Khalatnikov-Lifshitz universe. Phys. Rev. D **92**(12), 124068 (2015). https://doi.org/10.1103/PhysRevD.92.124068 (cit. on p. 99)
7. A. Bonanno, M. Reuter, Cosmology with self-adjusting vacuum energy density from a renormalization group fixed point. Phys. Lett. B **527**, 9–17 (2002). https://doi.org/10.1016/S0370-2693(01)01522-2. eprint: astro-ph/0106468 (cit. on p. 99)
8. A. Bonanno, A. Contillo, R. Percacci, Inflationary solutions in asymptotically safe f(R) theories. Class. Quantum Gravity **28**(14), 145026 (2011). https://doi.org/10.1088/0264-9381/28/14/145026. arXiv:1006.0192 [gr-qc] (cit. on p. 99)
9. A. Bonanno, An effective action for asymptotically safe gravity. Phys. Rev. D **85**(8), 081503 (2012). https://doi.org/10.1103/PhysRevD.85.081503. arXiv:1203.1962 [hep-th] (cit. on pp. 100, 103, 105, 107)
10. Planck Collaboration et al., Planck 2015 results. XIII. Cosmological parameters. A&A **594**, A13 (2016). https://doi.org/10.1051/0004-6361/201525830. arXiv:1502.01589 (cit. on pp. 100, 101, 104, 110, 111, 117)
11. A.A. Starobinsky, A new type of isotropic cosmological models without singularity. Phys. Lett. B **91**, 99–102 (1980). https://doi.org/10.1016/0370-2693(80)90670-X (cit. on pp. 100, 103, 105, 111)
12. D. Baumann, Inflation, in *Physics of the Large and the Small: TASI 2009*, ed. by C. Csaki, S. Dodelson (2011), pp. 523–686. https://doi.org/10.1142/9789814327183_0010 (cit. on pp. 100–102)
13. J. Martin, C. Ringeval, V. Vennin, Encyclopædia Inflationaris. Phys. Dark Universe **5**, 75–235 (2014). https://doi.org/10.1016/j.dark.2014.01.003. arXiv:1303.3787 (cit. on pp. 101, 102, 118)
14. A. de Felice, S. Tsujikawa, f(R) Theories. Living Rev. Relativ. **13**, 3 (2010). https://doi.org/10.12942/lrr-2010-3. arXiv:1002.4928 [gr-qc] (cit. on p. 102)
15. S. Capozziello, M. de Laurentis, Extended theories of gravity. Phys. Rep. **509**, 167–321 (2011). https://doi.org/10.1016/j.physrep.2011.09.003. arXiv:1108.6266 [gr-qc] (cit. on p. 102)
16. D. Benedetti, On the number of relevant operators in asymptotically safe gravity. EPL (Europhys. Lett.) **102**, 20007 (2013). https://doi.org/10.1209/0295-5075/102/20007. arXiv:1301.4422 [hep-th] (cit. on p. 104)

17. S. Coleman, E. Weinberg, Radiative corrections as the origin of spontaneous symmetry breaking. Phys. Rev. D **7**, 1888–1910 (1973). https://doi.org/10.1103/PhysRevD.7.1888 (cit. on p. 104)

18. H. Pagels, E. Tomboulis, Vacuum of the quantum Yang-Mills theory and magnetostatics. Nucl. Phys. B **143**, 485–502 (1978). https://doi.org/10.1016/0550-3213(78)90065-2 (cit. on p. 105)

19. A.B. Migdal, Vacuum polarization in strong non-homogeneous fields. Nucl. Phys. B **52**, 483–505 (1973). https://doi.org/10.1016/0550-3213(73)90575-0 (cit. on p. 105)

20. D.J. Gross, F. Wilczek, Asymptotically free gauge theories. I. Phys. Rev. D **8**, 3633–3652 (1973). https://doi.org/10.1103/PhysRevD.8.3633 (cit. on p. 105)

21. S.G. Matinyan, G.K. Savvidy, Vacuum polarization induced by the intense gauge field. Nucl. Phys. B **134**, 539–545 (1978). https://doi.org/10.1016/0550-3213(78)90463-7 (cit. on p. 105)

22. S.L. Adler, Short-distance perturbation theory for the leading logarithm models. Nucl. Phys. B **217**, 381–394 (1983). https://doi.org/10.1016/0550-3213(83)90153-0 (cit. on p. 105)

23. F. Fayos, R. Torres, A quantum improvement to the gravitational collapse of radiating stars. Class. Quantum Gravity **28**(10), 105004 (2011). https://doi.org/10.1088/0264-9381/28/10/105004 (cit. on p. 105)

24. A. Bonanno, M. Reuter, Renormalization group improved black hole spacetimes. Phys. Rev. D **62**(4), 043008 (2000). https://doi.org/10.1103/PhysRevD.62.043008. eprint: hep-th/0002196 (cit. on pp. 105, 106)

25. M. Reuter, H. Weyer, Quantum gravity at astrophysical distances? JCAP **12**, 001 (2004). https://doi.org/10.1088/1475-7516/2004/12/001. arXiv:hep-th/0410119 (cit. on p. 105)

26. M. Reuter, H. Weyer, Running Newton constant, improved gravitational actions, and galaxy rotation curves. Phys. Rev. D **70**(12), 124028 (2004). https://doi.org/10.1103/PhysRevD.70.124028. eprint: hep-th/0410117 (cit. on p. 105)

27. M. Reuter, H. Weyer, Renormalization group improved gravitational actions: a Brans-Dicke approach. Phys. Rev. D **69**(10), 104022 (2004). https://doi.org/10.1103/PhysRevD.69.104022. eprint: hep-th/0311196 (cit. on p. 105)

28. J.A. Dietz, T.R. Morris, Asymptotic safety in the f(R) approximation. J. High Energy Phys. **1**, 108 (2013). https://doi.org/10.1007/JHEP01(2013)108. arXiv:1211.0955 [hep-th] (cit. on p. 105)

29. M. Hindmarsh, I.D. Saltas, f(R) gravity from the renormalization group. Phys. Rev. D **86**(6), 064029 (2012). https://doi.org/10.1103/PhysRevD.86.064029 (cit. on pp. 105, 107)

30. E.J. Copeland, C. Rahmede, I.D. Saltas, Asymptotically safe Starobinsky inflation. Phys. Rev. D **91**(10), 103530 (2015). https://doi.org/10.1103/PhysRevD.91.103530 (cit. on pp. 105, 107)

31. O. Lauscher, M. Reuter, Flow equation of Quantum Einstein Gravity in a higher-derivative truncation. Phys. Rev. D **66**(2), 025026 (2002). https://doi.org/10.1103/PhysRevD.66.025026. eprint: hep-th/0205062 (cit. on pp. 105, 106)

32. S. Rechenberger, F. Saueressig, R^2 phase diagram of Quantum Einstein Gravity and its spectral dimension. Phys. Rev. D **86**(2), 024018 (2012). https://doi.org/10.1103/PhysRevD.86.024018. arXiv:1206.0657 [hep-th] (cit. on pp. 105, 106)

33. A. Codello, R. Percacci, C. Rahmede, Investigating the ultraviolet properties of gravity with a Wilsonian renormalization group equation. Ann. Phys. **324**, 414–469 (2009). https://doi.org/10.1016/j.aop.2008.08.008. arXiv:0805.2909 [hep-th] (cit. on pp. 106, 108)

34. K. Falls et al., Further evidence for asymptotic safety of quantum gravity. Phys. Rev. D **93**(10), 104022 (2016). https://doi.org/10.1103/PhysRevD.93.104022 (cit. on p. 107)

35. N.N. Bogolyubov, N.M. Krylov, *Introduction to Non-Linear Mechanics* (Princeton University Press, 1947). ISBN: 9780691079851 (cit. on p. 114)

36. N.N. Bogolyubov, Y.A. Mitropolski, *Asymptotic Methods in the Theory of Non-Linear Oscillations* (Gordon and Breach, 1961). ISBN: 978-0-677-20050-7 (cit. on p. 114)

37. P. Donà, A. Eichhorn, R. Percacci, Matter matters in asymptotically safe quantum gravity. Phys. Rev. D **89**(8), 084035 (2014). https://doi.org/10.1103/PhysRevD.89.084035. arXiv:1311.2898 [hep-th] (cit. on p. 118)

38. J. Meibohm, J.M. Pawlowski, M. Reichert, Asymptotic safety of gravity matter systems. Phys. Rev. D **93**(8), 084035 (2016). https://doi.org/10.1103/PhysRevD.93.084035. arXiv:1510.07018 [hep-th] (cit. on p. 118)

Chapter 6
Quantum Black Holes and Spacetime Singularities

The existence of spacetime singularities indicates the failure of the classical description of gravity within the framework of General Relativity. Such singularities typically emerge at the beginning of the universe and at the endpoint of gravitational collapse. In these extreme situations the evolution of causal geodesics across the singularity is not uniquely determined, i.e. the spacetime becomes "geodesically incomplete" [1, 2], and General Relativity loses its predictive power as a classical, deterministic theory.

Spacetime singularities are a rather general feature of General Relativity [3]. According to the still unproven Cosmic Censorship Conjecture (CCC) all singularities in physically realistic spacetimes must be covered by an event horizon [4], so that no signal from the singularity can reach the future null infinity. At variance of the cosmological singularity at the "origin of time", a black hole singularity at $r = 0$ develops in time because the radial coordinate is time-like inside of the black hole nucleus. In the latter case, both the singularity and the event horizon are dynamically produced by the inflowing material from the collapsing star and the backscattered gravitational waves. Although the structure of the interior solution is rather uncertain, most of the classical models describing the gravitational collapse contradict the Cosmic Censorship Hypothesis. Depending on the initial conditions, they can give rise to *naked singularities* (see [5] for a complete review). Leaving aside the possibility of fine-tuning the initial data to avoid the formation of naked singularities, one can argue that a consistent theory of Quantum Gravity may provide a solution

This chapter is based on the following publications:

• A. Bonanno, B. Koch, A. Platania-*Cosmic Censorship in Quantum Einstein Gravity*-Class. Quantum Grav. 34 (2017) 095012 [arXiv:1610.05299]

• A. Bonanno, B. Koch, A. Platania-*Asymptotically Safe gravitational collapse: Kuroda-Papapetrou RG-improved model*-PoS(CORFU2016)058

• A. Bonanno, B. Koch, A. Platania-*Gravitational collapse in Quantum Einstein Gravity*. Found. Phys. (2018). https://doi.org/10.1007/s10701-018-0195-7.

© Springer Nature Switzerland AG 2018
A. B. Platania, *Asymptotically Safe Gravity*, Springer Theses,
https://doi.org/10.1007/978-3-319-98794-1_6

to this problem: when the curvature reaches Planckian values quantum fluctuations of the geometry may deform the black hole interior and eventually halt the collapse. An interesting possibility, proposed in the context of minisuperspace models [6, 7] coupled to matter fields, is a transition to a de Sitter core of Planckian size [8, 9], a scenario which recently emerged also in the framework of Asymptotically Safe Gravity [10, 11]. The study of quantum-corrected black holes has been carried out in [12, 13]. Furthermore, recent studies have modeled the collapse in terms of an homogeneous interior surrounded by an RG-improved Schwarzschild exterior [14, 15].

In this final chapter we study the process of black hole formation by linking the dynamics of the gravitational collapse to the renormalization group flow evolution. In this description quantum gravitational effects emerge dynamically and alter the way the event horizon (EH) forms during the collapse. Subsequently, we will discuss the structure of the singularity resulting from the quantum-corrected Vaidya–Kuroda–Papapetrou (VKP) model [16–18]. Although the antiscreening character of the gravitational interaction favors the formation of naked singularities, it will be shown that Quantum Gravity fluctuations turn the classical space-like singularity into a gravitationally weak singularity [19] which remains naked for a finite amount of advanced time [20–22].

6.1   Vaidya–Kuroda–Papapetrou Spacetimes

In this section we summarize the relevant material on the VKP model [16–18] for the gravitational collapse. This model provides a simple description of the gravitational collapse and it was one of the first counterexamples to the Cosmic Censorship Conjecture.

In order to describe the spacetime structure around a spherically symmetric star of mass m, it is useful to introduce the ingoing Eddington–Finkelstein coordinates, where the Schwarzschild time coordinate is replaced by the advanced time parameter $v = t + r^*$, with

$$r^* \equiv r + r_s \log \left| \frac{r}{r_s} - 1 \right| \ , \tag{6.1}$$

where $r_s = 2mG_0$ is the Schwarzschild radius, $G_0 \equiv G_N$ being the Newton's constant.

With the aim of describing the dynamics of the gravitational collapse, we allow the mass of the black hole to be a function of the advanced time v. The resulting line element can thus be cast in the form

$$ds^2 = -f(r, v) \, dv^2 + 2 \, dv \, dr + r^2 \, d\Omega^2 \ , \tag{6.2}$$

with the lapse function $f(r, v)$ given by

$$f(r, v) = 1 - \frac{2 G_0 \, m(v)}{r} \quad . \tag{6.3}$$

This metric, defining the so-called Vaidya spacetimes [23], is an exact solution of the Einstein field equations in the presence of a Type II fluid with energy density [16, 23]

$$\rho(r, v) = \frac{\dot{m}(v)}{4\pi r^2} \quad . \tag{6.4}$$

It describes the spacetime surrounding a spherically symmetric object with variable mass $m(v)$.

The metric defined by the lapse function (6.3) is part of a larger class of metrics, known as generalized Vaidya spacetimes. This family of metrics, first introduced by Wang and Wu [24], is characterized by a lapse function

$$f(r, v) = 1 - \frac{2 M(r, v)}{r} \quad , \tag{6.5}$$

where the *generalized mass function* $M(r, v)$ depends on both the advanced time and the radial coordinate. The generalized Vaidya spacetimes are obtained as solutions of the Einstein field equations in the presence of a mixture of Type-I and Type-II fluids. The corresponding stress-energy tensor reads [25, 26]

$$T_{\mu\nu} = \underbrace{\rho l_\mu l_\nu}_{\text{Type II}} + \underbrace{(\sigma + p)(l_\mu n_\nu + l_\nu n_\mu) + p g_{\mu\nu}}_{\text{Type I}} \tag{6.6}$$

where $n_\mu l^\mu = -1$, $l_\mu l^\mu = 0$ and

$$\rho(r, v) = \frac{1}{4\pi G_0 r^2} \frac{\partial M(r, v)}{\partial v} \quad , \tag{6.7}$$

$$\sigma(r, v) = \frac{1}{4\pi G_0 r^2} \frac{\partial M(r, v)}{\partial r} \quad , \tag{6.8}$$

$$p(r, v) = -\frac{1}{8\pi G_0 r} \frac{\partial^2 M(r, v)}{\partial r^2} \quad . \tag{6.9}$$

In the standard Vaidya spacetime $M(r, v) \equiv G_0 \, m(v)$, so that the stress-energy tensor (6.6) reduces to the pure Type II one (radiation).

The gravitational collapse can be modeled as the implosion of a series of radiation shells, represented by the world lines $v = \text{const}$. The Vaidya spacetime can thus be used to describe an astrophysical object of growing mass $m(v)$. In the Vaidya–Kuroda–Papapetrou (VKP) model [16–18] this mass function is parametrized as follows

$$m(v) = \begin{cases} 0 & v < 0 \\ \lambda v & 0 \le v < \bar{v} \\ \bar{m} & v \ge \bar{v} \end{cases} \tag{6.10}$$

The spacetime is initially flat and empty (Minkowski spacetime). At $v = 0$ the radiation shells emanating from the collapsing star are injected and focused towards the origin $r = 0$, causing the mass of the central object to grow as $m(v) = \lambda v$. The VKP spacetime thus develops a persistent central singularity. The formation and evolution of the event horizon depend upon the details of the collapse dynamics. Finally, in $v = \bar{v}$ the collapse ends and the metric reduces to the Schwarzschild static solution, with final mass $\bar{m} = \lambda \bar{v}$.

The outcome of the collapse can be studied by solving the geodesic equation for null outgoing light rays, as it gives information on the causal structure of the spacetime. In the Vaidya geometry outgoing radial light rays are represented as solutions of

$$\frac{dr}{dv} = \frac{1}{2}\left(1 - \frac{2G_0 m(v)}{r}\right) . \tag{6.11}$$

The general solution to Eq. (6.11) is implicitly defined by

$$-\frac{2\,\mathrm{ArcTan}\left[\frac{v - 4r(v)}{v\sqrt{-1 + 16\lambda G_0}}\right]}{\sqrt{-1 + 16\lambda G_0}} + \log\left[2\lambda G_0 v^2 - r(v)\,v + 2\,r(v)^2\right] = C \tag{6.12}$$

where C is an arbitrary integration constant. A representative set of these solutions is depicted in Fig. 6.1. The classical VKP model is characterized by a critical value of the radiation rate, $\lambda_c \equiv 1/16G_0$, below which the singularity is *globally naked*. In this case the family of solutions in Eq. (6.11) reduces to the following implicit equation [27–29]

$$\frac{|r(v) - \mu_- v|^{\mu_-}}{|r(v) - \mu_+ v|^{\mu_+}} = \tilde{C} \tag{6.13}$$

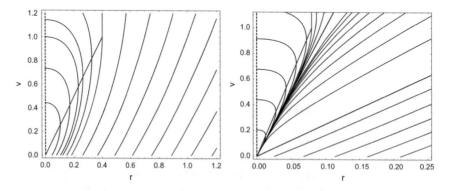

Fig. 6.1 Phase diagrams in the (r, v)–plane for the classical VKP model. The blue line represents the event horizon (EH), the purple line is the apparent horizon (AH), and the black curves are a sample of solutions to the classical geodesic equation. Left panel: For $\lambda > \frac{1}{16G_0}$ the singularity in $r = 0$ is covered by the event horizon. Right panel: In the case $\lambda \leq \frac{1}{16G_0}$ a Cauchy horizon $r_+(v) = \mu_+ v$ (external bold black line) is formed, and the singularity is naked

where $\widetilde{C}$ is a complex constant and

$$\mu_{\pm} = \frac{1 \pm \sqrt{1 - 16\,\lambda\,G_0}}{4} \ . \tag{6.14}$$

The above implicit equation, Eq. (6.13), has two simple linear solutions $r_{\pm}(v) = \mu_{\pm}\,v$, with $\mu_{\pm}$ defined in (6.14). The line $r_{-}(v) = \mu_{-}\,v$ is the tangent to the event horizon at $(r = 0,\, v = 0)$, while $r_{+}(v) = \mu_{+}\,v$ is the Cauchy horizon. The latter represents the first light ray emitted from the naked singularity. In particular all geodesics lying between these two linear solutions in the $(r,\,v)$-plane are light rays starting from the singularity $(r = 0,\, v = 0)$ and reaching the observer at infinity. Therefore if the initial conditions of the physical system entails $\lambda \le \lambda_c$, the gravitational collapse gives rise to a naked singularity. In this case the Cosmic Censor Conjecture is violated in its weak formulation.

6.2 Singularity Structure in Generalized Vaidya Spacetimes

The most severe problem related to the existence of singular spacetimes is the impossibility of uniquely determining the evolution of the spacetime beyond the singularity. According to the singularity theorems the existence of singularities in the solutions of Einstein field equations is quite general. However, such theorems do not specify the properties of such singularities, in particular their "nature".

The physical relevance of a singularity is determined by its strength [5, 25]. Following the Tipler classification [19], a singularity is "strong" if an object falling into the singularity is destroyed by the gravitational tidal forces, thus disappearing from the spacetime once the singularity is reached. According to [19], only strong curvature singularities are physically relevant. On the contrary the gravitationally weak or "integrable" singularities are considered as less severe, as the spacetime may be continuously extended across the singularity (see also [26] for an extensive discussion).

The singularity strength is determined by the behavior of light-like geodesics in the vicinity of the singularity [25]. Let us consider a generalized Vaidya spacetime with mass function $M(r, v)$. The geodesic equation for null rays is conveniently cast in the form of a system of coupled first order differential equations

$$\begin{cases} \frac{dv(t)}{dt} = N(r, v) \equiv 2\,r \\ \frac{dr(t)}{dt} = D(r, v) \equiv r - 2\,M(r, v) \ . \end{cases} \tag{6.15}$$

The fixed point solutions of the system (6.15) are identified by the conditions $r = 0$ and $M(0, v) = 0$, and define the singular loci of the generalized Vaidya spacetime [25]. Linearizing the system around a fixed point solution $(r_{\text{FP}}, v_{\text{FP}})$ yields

$$\begin{cases} \frac{dv(t)}{dt} = \dot{N}_{FP} \left(v - v_{FP}\right) + N'_{FP} \left(r - r_{FP}\right) \\ \frac{dr(t)}{dt} = \dot{D}_{FP} \left(v - v_{FP}\right) + D'_{FP} \left(r - r_{FP}\right) \end{cases} \tag{6.16}$$

where a "prime" denotes differentiation respect to r, a "dot" stands for differentiation with respect to v, and the subscript FP means that the derivatives are evaluated at the fixed point. The linearized system (6.16) defines the Jacobian matrix J

$$J \equiv \begin{pmatrix} \dot{N}_{FP} & N'_{FP} \\ \dot{D}_{FP} & D'_{FP} \end{pmatrix} . \tag{6.17}$$

In this description the behavior of the light-like geodesics in the proximity of the fixed point is completely determined by the eigenvalues of J

$$\gamma_{\pm} = \frac{1}{2} \left(\mathrm{Tr}\, J \pm \sqrt{(\mathrm{Tr}\, J)^2 - 4 \det J} \right) , \tag{6.18}$$

where

$$\mathrm{Tr}\, J = \dot{N}_{FP} + D'_{FP} \equiv 1 - 2\, M'_{FP} \tag{6.19a}$$

$$\det J = \dot{N}_{FP} D'_{FP} - \dot{D}_{FP} N'_{FP} \equiv 4\, \dot{M}_{FP} , \tag{6.19b}$$

and by the corresponding eigendirections. The latter define the characteristic lines

$$r_{\pm}(v) = r_{FP} + \frac{\gamma_{\pm}}{2} \left(v - v_{FP}\right) \tag{6.20}$$

passing through the fixed point (r_{FP}, v_{FP}). The slope of these lines, Eq. (6.20), determines the way radial null geodesics approach the fixed point. This description allows to locally characterize the singularity and determine its strength.

The central singularity is *locally naked* if there exists at least one light-like geodesic starting from the hypersurface $r = 0$ with a well defined tangent vector $\mathbf{v}$, and reaching the future null infinity. In terms of the above description, this condition is realized when the fixed point FP is a *repulsive node* ($\mathrm{Tr}\, J > 0$, $\det J > 0$, and $(\mathrm{Tr}\, J)^2 - 4 \det J > 0$). In the latter case the slope X_{FP} of the tangent vector $\mathbf{v}$

$$X_{FP} \equiv \lim_{(r,v) \to FP} \frac{v(r)}{r} \tag{6.21}$$

is determined by the non-marginal eigendirection (6.20) tangent to light-like geodesics at the singularity. Denoting by $\bar{\gamma}$ the eigenvalue (6.18) associated to this eigendirection, one easily finds $X_{FP} = 2/\bar{\gamma}$. Moreover, following [25], the parameter

$$S = \frac{X_{FP}^2}{2} \left(\partial_v M\right)_{FP} \tag{6.22}$$

is a measure of the singularity strength. A strong curvature singularity is thereby identified by the condition $S > 0$. On the contrary, if $S \leq 0$, the singularity is gravitationally weak (or integrable). When applying this analysis to the classical VKP model, one finds that the central singularity, represented by the fixed point $(0, 0)$ in the (r, v)–plane, is always strong [25]. Moreover the classical critical value $\lambda_c = 1/16G_0$ already emerges at the level of the linearized system: when $\lambda > 1/16G_0$ the fixed point is a spiral node, otherwise it is a pure repulsive node and corresponds to a locally naked singularity [27]. As we shall see, in the RG-improved case the critical value can only be found through the analysis of full geodesic equation.

6.3 RG-improved VKP Spacetimes

In this section we employ a RG improvement procedure to study the Quantum Gravity effects in the gravitational collapse of a massive star. Following [10], we start from the classical VKP solution (6.2) and perform the RG improvement by replacing G_0 with the running Newton's coupling G_k, Eq. (5.14). The RG-improved lapse function thus reads

$$f_1(r, v) = 1 - \frac{2\,m(v)}{r}\,\frac{G_0}{1 + \omega\,G_0\,k^2}\,. \tag{6.23}$$

In contrast to the method employed in Chap. 5, here the RG improvement is applied to a particular solution of the classical field equations. In this way the inclusion of quantum corrections directly modifies the spacetime geometry. Therefore, this strategy is appropriate for studying quantum-corrected black holes solutions [10, 11, 30–32].

A consistent description of the RG-improved gravitational collapse requires to find a relation between the renormalization group scale k and the actual collapse dynamics. Since the formation of realistic black holes is caused by the gravitational collapse of both matter and radiation, one can argue that the energy density of the collapsing fluid may serve as a physical infrared cutoff. Therefore, for actual calculations, we shall use the following scale-setting

$$k \equiv \xi \sqrt[4]{\rho} \tag{6.24}$$

where ξ is an arbitrary positive constant and the specific functional form $k(\rho)$ is dictated by simple dimensional arguments. Moreover, this is the only possible choice compatible with a conformally invariant theory at the NGFP, as gravity is supposed to be. The introduction of a generic functional form $k(\rho)$ would imply the presence of other mass scales not allowed at the NGFP.

In order to derive a quantum-corrected VKP spacetime, we shall first RG-improve the classical Vaidya solution with the cutoff identification (6.24) and ρ given by the energy density of the classical (bare) inflowing radiation, Eq. (6.4). Subsequently, we will compute the quantum-corrected stress-energy tensor arising from the RG-improved VKP metric. The infrared cutoff k is thus given by

$$k(r, v) \equiv \xi \sqrt[4]{\frac{\dot{m}(v)}{4\pi r^2}} \ . \tag{6.25}$$

Therefore, the quantum-corrected lapse function reads

$$f_1(r, v) = 1 - \frac{2 \, m(v) \, G_0}{r + \alpha \sqrt{\lambda}} \ , \qquad \alpha = \frac{\omega \, \xi^2 \, G_0}{\sqrt{4\pi}} \tag{6.26}$$

where $m(v)$ is the mass function introduced in Eq. (6.10). In contrast to the classical case, the RG-improved lapse function $f_1(r, v)$ does not diverge at $r = 0$

$$\lim_{r \to 0} f_1(r, v) = 1 - \frac{\sqrt{16\pi\lambda}}{\omega \, \xi^2} \, v \ . \tag{6.27}$$

The RG-improved VKP metric can be seen as a generalized Vaidya spacetime with generalized mass function

$$M_I(r, v) = G(r) \, m(v) = \frac{G_0 \, r}{r + \alpha \sqrt{\lambda}} \, m(v) \tag{6.28}$$

and corresponds to a non-trivial fluids mixture. In particular, when the collapse ends, the RG-improved VKP model defined by Eq. (6.26) reduces to a quantum-corrected Schwarzschild metric with lapse function

$$f_S(r, v) = 1 - \frac{2 \, \bar{m} \, G_0}{r + \alpha \sqrt{\lambda}} \ . \tag{6.29}$$

As expected, the continuity of the RG-improved mass function $M_I(r, v)$ along the $v = \bar{v}$ light-cone implies that the quantum-corrected spacetime does not converge to "pure" Schwarzschild, but only approaches it asymptotically (for $r \to \infty$).

Here it is important to notice that the event horizon (EH), separating the light rays converging back to singularity to the ones diverging to future null infinity, does not match the zeros of the lapse function, which instead define the so-called apparent horizon (AH). In the classical case the AH is given by the equation $r_{AH}(v) = 2 \, m(v) \, G_0$, while in the RG-improved Vaidya–Kuroda–Papapetrou model it reads

$$r_{AH}(v) = 2 \, m(v) \, G_0 - \alpha \sqrt{\lambda} = 2 \, m(v) \, G_0 - \frac{G_0 \, \xi^2}{g_*} \sqrt{\frac{\lambda}{4\pi}} \ , \tag{6.30}$$

with the condition $r_{AH}(v) \geq 0$. When the gravitational collapse ends, the Schwarzschild solution must be recovered and therefore the apparent and event horizons must converge to the RG-improved Schwarzschild radius $r_S = 2 \, \bar{m} \, G_0 - \alpha \sqrt{\lambda}$. The final mass $\bar{m}$ of the black hole, the value of $\bar{v}$, and the radiation rate λ are thus related by the condition $\bar{m} = \lambda \bar{v}$. In contrast to the classical case, the amount of advanced time $\bar{v}$ necessary to form a Schwarzschild black hole of radius $r_S \geq 0$ has a minimum value

$$r_S = 2\,\lambda\bar{v}\,G_0 - \alpha\sqrt{\lambda} \geq 0 \qquad \Rightarrow \qquad \bar{v} \geq v_{min}(\lambda) \equiv \frac{\xi^2}{2\,g_*}\sqrt{\frac{1}{4\pi\lambda}} \tag{6.31}$$

where the minimum v_{min} is a function of the radiation rate λ.

At last we have to remark that although the metric is regular at $r = 0$, the hypersurface $r = 0$ is actually singular. In fact in the quantum-corrected VKP model, assuming $m(v) = \lambda v$, the Ricci scalar R and Kretschmann scalar $K_r = R_{\alpha\beta\gamma\delta}R^{\alpha\beta\gamma\delta}$ read

$$R = \frac{4\alpha^2 m(v)\dot{m}(v)}{r^2\big(r + \alpha\sqrt{\dot{m}(v)}\,\big)^3} \simeq -\frac{G_0\sqrt{\lambda}\,v}{\alpha r^2} \; , \tag{6.32a}$$

$$K_r = \frac{16\,m(v)^2\left(r^4 + \big(r + \alpha\sqrt{\dot{m}(v)}\big)^2 r^2 + \big(r + \alpha\sqrt{\dot{m}(v)}\big)^4\right)}{r^4\big(r + \alpha\sqrt{\dot{m}(v)}\big)^6} \simeq \frac{16 G_0\sqrt{\lambda}\,v}{\alpha^2 r^4} \; , \tag{6.32b}$$

and diverge as $r \to 0$. Nevertheless, it should be noticed that the $r \to 0$ behavior is less singular than the classical case, in which $K_r \sim 1/r^6$.

6.3.1 *Gravitational Collapse in the RG-improved VKP Spacetime*

The causal structure of the RG-improved spacetime can be studied in the standard way, namely by analyzing the solutions to the geodesic equation for outgoing radial null rays. The RG-improved version of the geodesic equation reads

$$\dot{r}(v) = \frac{1}{2}\left(1 - \frac{2\,m(v)\,G_0}{r(v) + \alpha\,\sqrt{\lambda}}\right) \; . \tag{6.33}$$

The running of the Newton's coupling basically results in a shift of the radial coordinate $r(v)$ by $\alpha\,\sqrt{\lambda}$. The general solution to Eq. (6.33) can be analytically determined and is implicitly defined by

$$-\frac{2\,\text{ArcTan}\left[\frac{v - 4\,(r(v)+\alpha\sqrt{\lambda})}{v\sqrt{-1+16\,\lambda\,G_0}}\right]}{\sqrt{-1 + 16\,\lambda\,G_0}} + \log\!\left[2\lambda G_0 v^2 - (r(v) + \alpha\sqrt{\lambda})\,v + 2\,(r(v) + \alpha\sqrt{\lambda})^2\right] = C \tag{6.34}$$

where C is an integration constant.

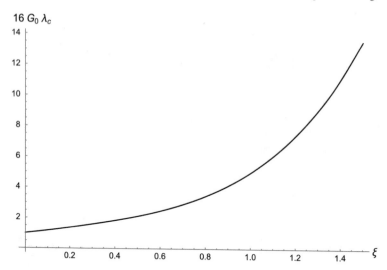

Fig. 6.2 Relative critical value $16G_0\lambda_c$ as a function of ξ. As it is clear from the picture, λ_c increases monotonically with ξ and in particular it reduces to the classical value when $\xi = 0$

In particular, in the case $\lambda \leq \frac{1}{16G_0}$ the above general solution reduces to the following implicit equation

$$\frac{|r(v) + \alpha\sqrt{\lambda} - \mu_- v|^{\mu_-}}{|r(v) + \alpha\sqrt{\lambda} - \mu_+ v|^{\mu_+}} = \tilde{C} \ . \tag{6.35}$$

This equation has two linear solutions

$$r_\pm(v) = -\alpha\sqrt{\lambda} + \mu_\pm v, \tag{6.36}$$

where $\mu_\pm$ are the parameters introduced in Eq. (6.14). In the case at hand the critical value λ_c cannot be analytically determined. The result of the numerical evaluation is summarized in Fig. 6.2, where the relative critical value $16G_0\lambda_c$ is plotted against the parameter ξ. This function monotonically increases with ξ and has an absolute minimum in $\xi = 0$, where the classical critical value is recovered. Therefore, the Quantum Gravity effects near the singularity causes an increase of the critical value λ_c, thus favoring the occurrence of naked singularities. However, as it will be discussed in the next section, the effect of the running Newton's constant vanishing in the ultraviolet limit is to render the singularity at $r = 0$ much milder and integrable.

The global behavior of light-like geodesics depends on the radiation rate λ. The phase diagram resulting from the numerical integration of Eq. (6.33) is shown in Fig. 6.3 for $\lambda \leq \lambda_c$. In the latter case the singularity is globally naked and, moreover, the event horizon emerges from $r = 0$ well after the formation of the singularity, thus

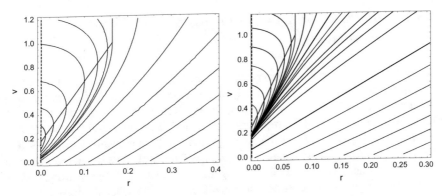

Fig. 6.3 Phase diagrams $(r, v(r))$ in the RG-improved VKP model, for $\frac{1}{16\,G_0} < \lambda \leq \lambda_c$ (left panel) and $\lambda_c \leq \frac{1}{16\,G_0}$ (right panel). The blue line is the EH, the purple line is the AH, and the black curves are particular solutions of the improved geodesic equation. For $\lambda \leq \lambda_c$ the singularity in $r = 0$ is globally naked, as the EH forms just after the formation of the singularity. Since in the improved case the linear solution $r_+(v)$ has no longer the meaning of a Cauchy horizon, the two cases depicted in these pictures are physically equivalents

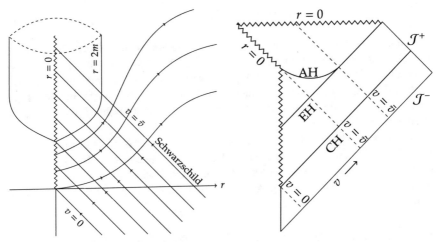

Fig. 6.4 Global structure of the spacetime for $\lambda \leq \lambda_c$. The singularity is globally naked and it extends in time, up to the formation of the event horizon for $v > 0$

allowing the singularity to be naked and persistent for a finite amount of advanced time v.

As a final remark, we note that the hypersurface $r = 0$ is time-like for $v < \tilde{v}$, null for $v = \tilde{v}$ and space-like for $v > \tilde{v}$, where $\tilde{v} = \frac{\alpha\sqrt{\lambda}}{2\lambda G_0}$ is defined by the limiting condition $\partial_\mu r \partial^\mu r \equiv f_1(r, v) = 0$. It coincides with the value of the advanced time v where the apparent horizon forms, $r_{AH}(\tilde{v}) = 0$ (see Fig. 6.4).

6.3.2   Singularity Structure in the RG-improved VKP Spacetime

We now apply the analysis described in Sect. 6.2 to study the singularity structure in the RG-improved VKP model. In this case the generalized mass function is decomposed as $M(r, v) = m(v) \, G(r)$, where the running Newton's constant $G(r)$ is obtained from Eq. (5.14) by replacing $k(r)$ with the infrared cutoff function (6.25), and results in Eq. (6.28). Since the running Newton's coupling vanishes for $r \to 0$, it follows that the fixed point condition $M(0, v) = 0$ holds at all values of the advanced time v. The singularity thus extends along the entire hypersurface $r = 0$ and defines a line of fixed points. In general a line of fixed points arises when the Jacobian determinant is zero, $\det J = 0$. In the case at hand the latter condition is satisfied due to the anti-screening behavior of the running Newton's constant in the high-energy regime

$$\det J \propto (\partial_v M)_{\mathrm{FP}} \propto \lim_{r \to 0} G(r) = 0 \tag{6.37}$$

and implies that

$$S \propto (\partial_v M)_{\mathrm{FP}} = 0 \;\; . \tag{6.38}$$

We can conclude that Quantum Gravity fluctuations near the singularity turn the strong curvature singularity of the classical Vaidya spacetime into a line of gravitationally weak singularities. Notably this result does not depend on the cutoff identification (6.25). In fact, provided that $k(r) \to \infty$ for $r \to 0$, the Newton's coupling vanishes for $r \to 0$ and, subsequently, the same holds for the strength parameter (6.22).

The above analysis shows that the linearized geodesic equation of the quantum-corrected VKP model gives rise to a line of integrable singularities. In order to understand whether these singularities are naked or not, a detailed study of the dynamical system (6.16) is needed. For a given fixed point situated at $P \equiv (0, v_0)$, Eq. (6.19) reads

$$\mathrm{Tr} \, J = 1 - 2 \, M'_{\mathrm{FP}} = 1 - \frac{2 \lambda v_0 \, G_0}{\alpha \sqrt{\lambda}} \qquad \det J = 4 \, \dot{M}_{\mathrm{FP}} = 0 \;\; . \tag{6.39}$$

Accordingly, the eigenvalue γ_- is always zero, while γ_+ depends on the particular fixed point $(0, v_0)$ in the singular line $r = 0$

$$\gamma_+(v_0) \equiv \mathrm{Tr} \, J = 1 - \frac{2 \lambda v_0 \, G_0}{\alpha \sqrt{\lambda}} \;\; . \tag{6.40}$$

In dynamical systems language, these peculiar features entail that the fixed points $(0, v_0)$ are *improper nodes*. The resulting line of improper nodes is thus characterized by one marginal direction $r = 0$ and by a family of non-marginal characteristic lines whose slope depends on the precise location of the fixed point

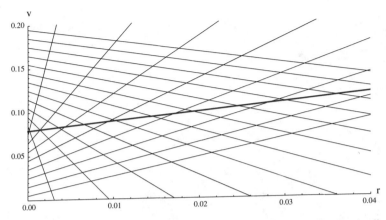

Fig. 6.5 Characteristic lines controlling the behavior of the system near the singular line, Eq. (6.41). The trajectories starting above (under) the AH (purple line) have negative (positive) slope. The qualitative behavior of these characteristic directions does not depend on the radiation rate λ

$$v = v_0 + \frac{2\,r}{\gamma_+(v_0)} \quad . \tag{6.41}$$

For a given fixed point P, the line in Eq. (6.41) is tangent to the geodesic ending in P. Depending on v_0 the slope of such lines can be either positive or negative. The value $\tilde{v}_0$ at which the slope inverts its sign is set by the condition

$$r_{\mathrm{AH}}(\tilde{v}_0) = 0 \quad . \tag{6.42}$$

Moreover, since $\gamma_+(v_0) \equiv \mathrm{Tr}\,J$, a positive (negative) slope of (6.41) in P implies that the fixed point is repulsive (attractive) along the corresponding non-marginal characteristic direction. The family of characteristic lines in Eq. (6.41) is shown in Fig. 6.5. It is worth noticing that these curves (black lines in the figure) cannot cross each other and the presence of intersections is due to the linearization of the geodesic equation around $r = 0$: moving away from the singularity the non-linear effects of the original dynamical system deform the characteristic lines so that different solutions never intersect each others. Note that in contrast to the classical case [25, 27], the qualitative behavior of the trajectories and the existence of a line of improper nodes do not depend on the precise value of the radiation rate λ. In particular no critical value appears. Accordingly, the singularity of the RG-improved model is never *locally naked*, independently of λ. On the other hand the analysis in Sect. 6.3 shows that there exists a critical value λ_c below which the singularity is *globally naked*. The apparent mismatch is due to the fact that the analytical solutions in Sect. 6.3 are obtained by solving the full geodesic equation, while the study of the singularity with the approach of [25] is performed by linearizing the system around $r = 0$. The critical value of the radiation rate λ_c is restored once the *full* RG-improved geodesic equation is considered, and must then be related to the classical collapse dynamics.

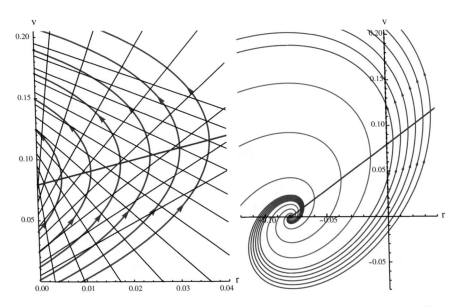

Fig. 6.6 Solutions of the full RG-improved geodesic equation for $\lambda > \lambda_c$. Left panel: the non-linear effects fold and join the characteristic lines (black lines), resulting in a set of characteristic curves (blue lines) linking couple of fixed points. Right panel: the line of fixed points and the corresponding characteristic curves can also be understood in terms of the "shifted classical solution": they are generated by the shifted spiral node characterizing the classical VKP model for $\lambda > \lambda_c$

When the non-linear effects are taken into account, the family of characteristic curves in Eq. (6.41) becomes a continuum set of heteroclinic orbits between couple of fixed points, as it is shown in the left panel of Fig. 6.6 for $\lambda > \lambda_c$.

As a final remark, the study of the full geodesic equation allows to interpret the line of fixed points and the corresponding v_0-dependent characteristic curves as result of a spiral node located in the unphysical part of the diagram, $r < 0$. In fact, the RG-improved geodesic equation can be obtained from the classical one by means of a shift in the radial coordinate $r(v) \rightarrow r(v) + \alpha\sqrt{\lambda}$. Since the classical VKP model has a singularity in $(0, 0)$, which is a spiral node for $\lambda > \lambda_c$ [27], the inclusion of the leading quantum effects moves this node to the point $(-\alpha\sqrt{\lambda}, 0)$. This (shifted) spiral node is the "source" which produces the line of fixed points $r = 0$ in the RG-improved system, as it is clear from the right panel of Fig. 6.6.

References

1. R.P. Geroch, Local characterization of singularities in general relativity. J. Math. Phys. **9**, 450–465 (1967). https://doi.org/10.1063/1.1664599. (cit. on p. 123)
2. R.P. Geroch, What is a singularity in general relativity? Ann. Phys. **48**, 526–540 (1968). https://doi.org/10.1016/0003-4916(68)90144-9. (cit. on p. 123)

3. S.W. Hawking, R. Penrose, The singularities of gravitational collapse and cosmology. Proc. Roy. Soc. Lond. Ser. A **314**, 529–548 (1970). https://doi.org/10.1098/rspa.1970.0021. (cit. on p. 123)

4. R. Penrose, *Gravitational Collapse: The Role of General Relativity* (Rivista del Nuovo Cimento 1, 1969), p. 252 (cit. on p. 123)

5. P.S. Joshi, *Gravitational Collapse and Spacetime Singularities* (Cambridge University Press, 2012). ISBN: 9781107405363 (cit. on pp. 124, 128)

6. J.B. Hartle, S.W. Hawking, Wave function of the universe. Phys. Rev. D **28**, 2960–2975 (1983). https://doi.org/10.1103/PhysRevD.28.2960. (cit. on p. 124)

7. A. Vilenkin, Quantum cosmology and the initial state of the universe. Phys. Rev. D **37**, 888–897 (1988). https://doi.org/10.1103/PhysRevD.37.888. (cit. on p. 124)

8. V.P. Frolov, M.A. Markov, V.F. Mukhanov, Through a black hole into a new universe? Phys. Lett. B **216**, 272–276 (1989). https://doi.org/10.1016/0370-2693(89)91114-3. (cit. on p. 124)

9. I. Dymnikova, Vacuum nonsingular black hole. Gen. Relativ. Gravit. **24**, 235–242 (1992). https://doi.org/10.1007/BF00760226 (cit. on p. 124)

10. A. Bonanno, M. Reuter, Renormalization group improved black hole spacetimes. Phys. Rev. D **62**(4), 043008 (2000). https://doi.org/10.1103/PhysRevD.62.043008. eprint: hep-th/0002196 (cit. on pp. 124, 130)

11. B. Koch, F. Saueressig, Structural aspects of asymptotically safe black holes. Class. Quantum Gravity **31**(1), 015006 (2014). https://doi.org/10.1088/0264-9381/31/1/015006. arXiv:1306.1546 [hep-th] (cit. on pp. 124, 130)

12. O. Lauscher, M. Reuter, Flow equation of Quantum Einstein Gravity in a higher-derivative truncation. Phys. Rev. D **66**(2), 025026 (2002). https://doi.org/10.1103/PhysRevD.66.025026. eprint: hep-th/0205062 (cit. on p. 124)

13. A. Codello, R. Percacci, Fixed points of higher-derivative gravity. Phys. Rev. Lett. **97**(22), 221301 (2006). https://doi.org/10.1103/PhysRevLett.97.221301. eprint: hep-th/0607128 (cit. on p. 124)

14. R. Torres, Singularity-free gravitational collapse and asymptotic safety. Phys. Lett. B **733**, 21–24 (2014). https://doi.org/10.1016/j.physletb.2014.04.010. arXiv:1404.7655 [gr-qc] (cit. on p. 124)

15. R. Torres, F. Fayos, Singularity free gravitational collapse in an effective dynamical quantum spacetime. Phys. Lett. B **733**, 169–175 (2014). https://doi.org/10.1016/j.physletb.2014.04.038. arXiv:1405.7922 [gr-qc] (cit. on p. 124)

16. P.C. Vaidya, An analytical solution for gravitational collapse with radiation. ApJ **144**, 943 (1966). https://doi.org/10.1086/148692 (cit. on pp. 124–126)

17. Y. Kuroda, Naked singularities in the Vaidya spacetime. Progr. Theor. Phys. **72**, 63–72 (1984). https://doi.org/10.1143/PTP.72.63. (cit. on pp. 124, 126)

18. A. Papapetrou, Formation of a singularity and causality, in *A Random Walk in Relativity and Cosmology*, ed. by M. Dadhich et al. (1985), pp. 184–191 (cit. on pp. 124, 126)

19. F.J. Tipler, On the nature of singularities in general relativity. Phys. Rev. D **15**, 942–945 (1977). https://doi.org/10.1103/PhysRevD.15.942. (cit. on pp. 124, 128)

20. A. Bonanno, B. Koch, A. Platania, Cosmic censorship in Quantum Einstein Gravity. Class. Quantum Gravity **34**(9), 095012 (2017). https://doi.org/10.1088/1361-6382/aa6788. arXiv:1610.05299 [gr-qc] (cit. on p. 124)

21. A. Bonanno, B. Koch, A. Platania, Asymptotically safe gravitational collapse: Kuroda-Papapetrou RG-improved model. In: *PoS CORFU2016* (2017), p. 058 (cit. on p. 124)

22. A. Bonanno, B. Koch, A. Platania, Gravitational collapse in Quantum Einstein Gravity. Found. Phys. (2018). https://doi.org/10.1007/s10701-018-0195-7

23. P.C. Vaidya, Nonstatic solutions of Einstein's field equations for spheres of fluids radiating energy. Phys. Rev. **83**, 10–17 (1951). https://doi.org/10.1103/PhysRev.83.10. (cit. on p. 125)

24. A. Wang, Y. Wu, Generalized Vaidya solutions. Gen. Relativ. Gravit. **31**, 107–114 (1999). https://doi.org/10.1023/A:1018819521971. eprint: gr-qc/9803038 (cit. on p. 125)

25. M.D. Mkenyeleye, R. Goswami, S.D. Maharaj, Gravitational collapse of generalized Vaidya spacetime. Phys. Rev. D **90**(6), 064034 (2014). https://doi.org/10.1103/PhysRevD.90.064034 (cit. on pp. 125, 128, 129, 137)

26. P.S. Joshi, *Global Aspects in Gravitation and Cosmology*, vol. 87 (Oxford UP (International series of monographs on physics), 1993) (cit. on pp. 125, 128)

27. W. Israel, The formation of black holes in nonspherical collapse and cosmic censorship. Can. J. Phys. **64**, 120–127 (1986). https://doi.org/10.1139/p86-018. (cit. on pp. 126, 129, 137, 138)

28. I.H. Dwivedi, P.S. Joshi, On the nature of naked singularities in Vaidya spacetimes. Class. Quantum Gravity **6**, 1599–1606 (1989). https://doi.org/10.1088/0264-9381/6/11/013. (cit. on p. 126)

29. I.H. Dwivedi, P.S. Joshi, On the nature of naked singularities in Vaidya spacetimes: II. Class. Quantum Gravity **8**(7), 1339–1348 (1991). https://doi.org/10.1088/0264-9381/8/7/010. (cit. on p. 126)

30. A. Bonanno, M. Reuter, Quantum gravity effects near the null black hole singularity. Phys. Rev. D **60**(8), 084011 (1999). https://doi.org/10.1103/PhysRevD.60.084011. eprint: gr-qc/9811026 (cit. on p. 130)

31. A. Bonanno, M. Reuter, Spacetime structure of an evaporating black hole in quantum gravity. Phys. Rev. D **73**(8), 083005 (2006). https://doi.org/10.1103/PhysRevD.73.083005. eprint: hep-th/0602159 (cit. on p. 130)

32. R. Torres, F. Fayos, On the quantum corrected gravitational collapse. Phys. Lett. B **747**, 245–250 (2015). https://doi.org/10.1016/j.physletb.2015.05.078. arXiv:1503.07407 [gr-qc] (cit. on p. 130)

Part IV
Conclusions

Chapter 7
Conclusions

The generalized notion of renormalizability, naturally arising from the Wilsonian Renormalization Group, is related to the existence of ultraviolet fixed points of the renormalization group flow. On this basis, gravity may result in a finite and predictive quantum theory if its flow converges to a Non-Gaussian Fixed Point (NGFP) in the ultraviolet limit. Such a non-trivial fixed point would guarantee the renormalizability of gravity and define its ultraviolet completion. The resulting Asymptotic Safety scenario for Quantum Gravity may provide a consistent description of the gravitational interaction from trans-Planckian to cosmological scales.

In this Ph.D. thesis we investigated two important aspects of Asymptotically Safe gravity. Firstly, we discussed the formulation of a Functional Renormalization Group Equation (FRGE) in the Arnowitt-Deser-Misner (ADM) formalism and we analyzed the gravitational renormalization group flow in the presence of an arbitrary number of matter fields. Secondly we studied the consequences of Asymptotic Safety in situations where quantum gravitational effects are expected to be important, namely in the very early universe and at the end-point of a gravitational collapse.

The thesis is organized in three parts. In Part 1 the main idea underlying the Wilsonian Renormalization Group is introduced. The definition of the Wilsonian action and the subsequent derivation of the Wegner-Houghton equation allowed us to introduce the generalized notion of renormalizability in a very intuitive and natural manner. As we have seen, the renormalizability of quantum field theories is based on the existence of fixed points which serve as ultraviolet attractors for the renormalization group flow and define the high-energy completion of the quantum theory. Subsequently, we introduced the Effective Average Action (EAA), which is a basic concept underlying modern Functional Renormalization Group (FRG) techniques, and we discussed its application to the case of gauge theories. Finally, we focused on the gravitational renormalization group flow projected onto the Einstein-Hilbert subspace and derived the beta functions for the cosmological constant and Newton's coupling within the metric approach to Asymptotic Safety. The resulting renormalization group flow in four spacetime dimensions has a single UV-attractive NGFP

© Springer Nature Switzerland AG 2018
A. B. Platania, *Asymptotically Safe Gravity*, Springer Theses,
https://doi.org/10.1007/978-3-319-98794-1_7

coming with a complex pair of critical exponents. Although we only reviewed the simple case of the Einstein-Hilbert truncation, more elaborate truncations confirm this picture [1–29].

In Part 2 we employed the ADM-formalism to derive the gravitational renormalization group flow in the presence of a foliation structure. In contrast to previous studies [30, 31], our construction does not require the time direction to be compact and also removes spurious singularities in the beta functions. The analytic continuation can be implemented by Wick-rotating all time-like quantities according to Eq. (4.10), provided that the background geometry possesses a time-like Killing vector. In our discussion we restricted our attention to an ADM-decomposed (Euclidean) Einstein-Hilbert action, supplemented by an arbitrary number of minimally coupled scalar, vector and Dirac fields. The renormalization group flow is then evaluated on a cosmological ($S^1 \times T^d$) Friedmann-Robertson-Walker background, so that the flow equations for the cosmological constant and Newton's coupling are encoded in the volume factor and *extrinsic* curvature terms constructed from the background. It is important to stress that the main features of the renormalization group flow on foliated spacetimes do not depend on the topology of the background geometry [32]. For instance, in the case of a ($S^1 \times S^d$) background the Newton's coupling is given by the coefficient multiplying the *intrinsic* curvature term. Nevertheless, the resulting fixed point structure and the main features of the flow diagram [32] match our findings. On the other hand, our construction is based on a cosmological background and, therefore, is perfectly suitable for the computation of correlation functions and power spectra arising in a cosmological context. In particular this formalism can be used to study the primordial gravitational wave spectrum, a problem we hope to address in a future work.

One of the peculiar features of our computation lies in a novel gauge-fixing scheme, which provides relativistic dispersion relations for all component fields involved in the calculation: all fields propagate with the same speed of light when the dispersion relations are evaluated on a Minkowski background. This condition fixes the gauge uniquely. In Sect. 4.2 we derived the beta functions encoding the running of the Newton's and cosmological couplings for a general $D = (d + 1)$-dimensional spacetime manifold and for an arbitrary number of matter fields, Eq. (4.49). In the case of pure gravity, the renormalization group equations give rise, in $D = 4$ spacetime dimensions, to a unique UV-attractive NGFP coming with a complex pair of critical exponents. This is the same behavior as the one observed in the metric formulation of Quantum Einstein Gravity (QEG) [1–29] and earlier work on the ADM-formalism [30, 31]. Flowing away from the NGFP, the renormalization group flow is dominated by the interplay of the NGFP, controlling the behavior of gravity for ultra-high energies, and the Gaussian Fixed Point (GFP), governing the long-distance regime. The corresponding phase diagram is depicted in Fig. 4.2. The latter matches the main features of the renormalization group flow encountered in the metric approach to Asymptotic Safety [4]. In particular, the renormalization group trajectories can be classified according to the infrared sign of the cosmological constant.

In Sect. 4.3.2 we discussed the fixed point structure arising in foliated gravity-matter systems. The latter is characterized by a rich family of fixed points and, depending on the number of scalars, vectors and fermions, the renormalization group flow can exhibit up to three fixed points. In particular, at the level of the beta functions (4.49), the effect of the matter fields can be encoded in two "deformation parameters" (d_g, d_λ) which depend on the number of matter fields in each sector. The fixed points for a specific gravity-matter model can then be determined by evaluating the map relating its field content to the deformation parameters. The most important finding in this regard is that the most commonly studied matter models, including the Standard Model (SM) of particle physics and its modifications, are located in areas of the (d_g, d_λ)–diagram which give rise to a unique UV-attractive fixed point with $\lambda_* < 0$, and positive and real critical exponents. These results are summarized in Table 4.4. A prototype of renormalization group flow arising from the latter family of gravity-matter models is shown in Fig. 4.5. In these cases the main features familiar from Asymptotically Safe gravity remain intact. The characteristic spiraling behavior of the renormalization group flow is lost, though, and the NGFP sits into the left part of the diagram, $\lambda_* < 0$. These findings complement earlier studies based on the metric formalism [33, 34] and provide a first indication that the asymptotic safety mechanism encountered in the case of pure gravity may carry over to the case of foliated gravity-matter systems with a realistic matter content. On the other hand, a comparison of Table 4.4 with the results obtained in [33] within the metric approach to Asymptotic Safety shows that these approaches give rise to slightly different results. For instance, according to [33], only the SM and its minor modifications are compatible with Asymptotic Safety, while the Minimal Supersymmetric Standard Model (MSSM) and Grand Unified Theories (GUTs) are excluded. While this discrepancy can be attributed to a different choice of the regulator, it is very likely that the coupling of gravity to matter fields may emphasize a substantial difference between the metric and foliated approaches to Asymptotic Safety. In fact, in the light of recent studies related to gravity in $D = 2$ dimensions [35], it is conceivable that the FRGE based on the metric formalism and the ADM-formalism actually access different universality classes for gravity and gravity-matter systems.

The final part of Chap. 4 studies the fixed point structure emerging from the renormalization group flow of gravity as a function of the spacetime dimension. The GFP is always present and its stability coefficients are given by the canonical scaling dimensions associated with the cosmological constant and Newton's coupling. In addition, we found two families of NGFPs whose most important properties are summarized in Fig. 4.7. In particular, for $2.37 \leq D \leq 3.25$ the renormalization group flow is characterized by a family of non-trivial saddle points and by a family of UV-attractive fixed points coming with real and positive critical exponents. For $D > 3.40$ the latter critical exponents become complex and give rise to the standard NGFP underlying the Asymptotic Safety of pure Quantum Gravity. Notably, these findings are in good agreement with the ones obtained within lattice Quantum Gravity [36] and through the discretized Wheeler-de Witt equation [37, 38]. Moreover, in $D = 4$ spacetime dimensions, the critical exponents are very similar to those obtained from foliated spacetimes using the Matsubara formalism [30]. The most interesting aspect emerging from this analysis is that the existence of a second non-trivial fixed point could render the behavior of gravity well defined for all energy scales, resolving the IR singularity usually encountered when only one NGFP is present. The interplay of these fixed points leads to a new long-distance modification of gravity in which both the Newton's coupling and the cosmological constant are dynamically driven to zero. This new phase is depicted in Fig. 4.9, while the phase diagram corresponding to this case is plotted in the left panel of Fig. 4.8. Building up on these observations, [32] showed that the same mechanism may also be realized in $D = 4$ dimensions. In this case, the transition between the classical phase and the new long-distance phase of gravity may be visible on cosmic scales.

In Part 3 we studied some applications of Asymptotically Safe Gravity in astrophysics and cosmology. Over the years several investigations in the context of Asymptotic Safety have shown how renormalization group improved cosmology can naturally describe the cosmological evolution of the universe from the initial singularity to the late time regime described by General Relativity. A way to incorporate the scale dependence of the gravitational couplings into the description of dynamical gravitational phenomena is the renormalization group improvement procedure. Using this method, we discussed a family of inflationary models emerging from the Quantum Gravity modifications due to the scaling of couplings around the NGFP (Chap. 5). In particular, we extended the analysis of [39] by taking into account the additional relevant direction associated to the R^2 operator. Assuming the critical exponents to be real numbers, we approximated the renormalization group flow in a way that allows an analytical study of the inflationary dynamics within the slow-roll approximation. We then restricted ourselves to the class of potentials providing a well definite exit from inflation, followed by the standard phase of parametric oscillations of the inflaton field. The resulting family of RG-improved inflationary potentials gives rise to values for the spectral index and tensor-to-scalar ratio in agreement with the recent Planck data on the power spectrum of Cosmic Microwave Background (CMB) anisotropies [40]. In particular, although our model is only a small modification of the well known Starobinsky model [41], it predicts values for the tensor-to-scalar ratio that are significantly higher. This feature makes the quantum-corrected Starobinsky model significantly different from the original one.

While the RG-improved Starobinsky model shows a very good agreement with the Planck data, the inclusion of matter fields may be important to correctly describe the physics of the very early universe. In particular, as we have seen in Chap. 4, the critical exponents underlying the behavior of gravity in the proximity of the NGFP strongly depend on the matter content of the theory, and their precise values are important for the description of the primordial evolution of the universe. When matter fields are taken into account within the Einstein-Hilbert truncation, the strategy employed to study the RG-improved cosmic inflation, combined with the Planck data on the power spectrum amplitude and our findings of Chap. 4 actually furnish important constraints on the primordial matter content of the universe. According to this analysis, the property of the power spectrum to be nearly scale-invariant indirectly constrains the second largest critical exponent to $\theta_2 \leq 4$. Moreover, the Planck data on the power spectrum amplitude suggest that the ultraviolet value of the cosmological constant should be negative. Notably, when including the Quantum Gravity corrections into the beta functions of the Standard Model couplings, a negative value of λ_* is required to define the ultraviolet completion for the Higgs-Yukawa couplings [42]. According to Table 4.4, typical models compatible with such constraints are the Standard Model, its minor modifications including a small number of extra fields, and the MSSM. On the contrary, GUT-type theories are disfavored by the analysis.

Finally, in Chap. 6, we studied the problem of the black hole formation by linking the dynamics of the gravitational collapse with the RG evolution of the Newton's coupling as predicted by Asymptotic Safety. Under the assumption of spherical symmetry, the gravitational collapse of a massive star can be modeled by means of a generalized Vaidya spacetime. Starting from the classical Vaidya-Kuroda-Papapetrou model, we used the RG improvement procedure to obtain the corresponding quantum-corrected spacetime. In particular, the inclusion of the leading Quantum Gravity effects resulted in an effective mass function, whose analytical form is completely determined by the running Newton's coupling. In order to study the outcome of the collapse, we analyzed the global solutions of the RG-improved geodesic equation for outgoing null rays. This analysis showed that the inclusion of the leading quantum-effects in the Vaidya-Kuroda-Papapetrou model is not enough to eliminate the central singularity. Moreover, the RG-improved model favors the formation of naked singularities. On the other hand, the anti-screening behavior of the Newton's coupling modifies the structure of the central singularity so that the classical strong singularity is turned into a line of integrable, gravitationally weak singularities. Therefore, in contrast to the classical case, the quantum-corrected Vaidya-Kuroda-Papapetrou spacetime can be continuously extended across the singularity. Possible generalizations of this model should clearly include the angular momentum. Moreover, it would be interesting to investigate possible astrophysical consequences of this class of models, in particular the possibility that Quantum Gravity effects may be detected from the signals emitted by integrable naked singularities. In fact, naked singularities are possible candidate to be considered real astrophysical objects [43]. Whether or not they are realized in nature depends on the initial conditions and dynamics of the collapse. The most interesting feature of this possibility is that, since every signal generated near a naked singularity can reach the future null infinity, these objects could provide a very interesting laboratory to study high-energy physics, potentially shedding light on extensions of the SM and Quantum Gravity.

An important limitation of our analyses lies in the simple truncation schemes and tensorial structures we employed throughout this work. The inclusion of other relevant operators could change the features of the renormalization group flow and, subsequently, may slightly modify the dynamics in the RG-improved models we considered. Most importantly, the RG improvement procedure allows to include the leading quantum corrections only, and should be seen as a first step towards the understanding of the high-energy modifications of gravitational phenomena. Nevertheless, we have seen that Quantum Gravity corrections play a crucial role in the description of the early universe cosmology, as well as in black holes physics. The Quantum Gravity modifications induced by Asymptotic Safety can in fact lead to interesting astrophysical consequences and cosmological scenarios. Moreover, the high-energy behavior of gravity is sensitive to the matter content of the universe and it turns out that SM-type theories are favored by this scenario. The inclusion of matter fields is therefore of fundamental importance for studying Quantum Gravity and its phenomenological implications. Although falsifying Quantum Gravity theories through phenomenology is still far from nowadays reach, the study of quantum gravitational effects and their implications can provide important constraints and open the possibility of finding astrophysical signatures of Quantum Gravity and, eventually, discriminate between different theories.

References

1. M. Reuter, Nonperturbative evolution equation for quantum gravity. Phys. Rev. D **57**, 971–985 (1998). https://doi.org/10.1103/PhysRevD.57.971. eprint: hep-th/9605030 (cit. on pp. 144, 145)
2. W. Souma, Non-trivial ultraviolet fixed point in quantum gravity. Prog. Theor. Phys. **102**, 181–195 (1999). https://doi.org/10.1143/PTP.102.181. eprint: hep-th/9907027 (cit. on pp. 144, 145)
3. O. Lauscher, M. Reuter, Ultraviolet fixed point and generalized flow equation of quantum gravity. Phys. Rev. D **65**(2), 025013 (2002). https://doi.org/10.1103/PhysRevD.65.025013. eprint: hep-th/0108040 (cit. on pp. 144, 145)
4. M. Reuter, F. Saueressig, Renormalization group flow of quantum gravity in the Einstein-Hilbert truncation. Phys. Rev. D **65**(6), 065016 (2002). https://doi.org/10.1103/PhysRevD.65.065016. eprint: hep-th/0110054 (cit. on pp. 144, 145)
5. D.F. Litim, Fixed points of quantum gravity. Phys. Rev. Lett. **92**(20), 201301 (2004). https://doi.org/10.1103/PhysRevLett.92.201301. eprint: hep-th/0312114 (cit. on pp. 144, 145)
6. O. Lauscher, M. Reuter, Flow equation of Quantum Einstein Gravity in a higher-derivative truncation. Phys. Rev. D **66**(2), 025026 (2002). https://doi.org/10.1103/PhysRevD.66.025026. eprint: hep-th/0205062 (cit. on pp. 144, 145)
7. A. Codello, R. Percacci, C. Rahmede, Ultraviolet properties of f(R)-gravity. Int. J. Mod. Phys. A **23**, 143–150 (2008). https://doi.org/10.1142/S0217751X08038135. arXiv:0705.1769 [hep-th] (cit. on pp. 144, 145)
8. P.F. Machado, F. Saueressig, On the renormalization group flow of f(R)-gravity. Phys. Rev. D **77**, 124045 (2008). https://doi.org/10.1103/PhysRevD.77.124045. eprint: arXiv:0712.0445 (cit. on pp. 144, 145)
9. A. Codello, R. Percacci, C. Rahmede, Investigating the ultraviolet properties of gravity with a Wilsonian renormalization group equation. Ann. Phys. **324**, 414–469 (2009). https://doi.org/10.1016/j.aop.2008.08.008. arXiv:0805.2909 [hep-th] (cit. on pp. 144, 145)
10. K. Falls et al., Further evidence for asymptotic safety of quantum gravity. Phys. Rev. D **93**(10), 104022 (2016). https://doi.org/10.1103/PhysRevD.93.104022. (cit. on pp. 144, 145)
11. M. Demmel, F. Saueressig, O. Zanusso, RG flows of quantum Einstein gravity in the linear-geometric approximation. Ann. Phys. **359**, 141–165 (2015). https://doi.org/10.1016/j.aop.2015.04.018. arXiv:1412.7207 [hep-th] (cit. on pp. 144, 145)
12. A. Codello, R. Percacci, Fixed points of higher-derivative gravity. Phys. Rev. Lett. **97**(22), 221301 (2006). https://doi.org/10.1103/PhysRevLett.97.221301. eprint: hep-th/0607128 (cit. on pp. 144, 145)
13. D. Benedetti, P.F. Machado, F. Saueressig, Asymptotic safety in higher-derivative gravity. Mod. Phys. Lett. A **24**, 2233–2241 (2009). https://doi.org/10.1142/S0217732309031521. arXiv:0901.2984 [hep-th] (cit. on pp. 144, 145)
14. D. Benedetti, P.F. Machado, F. Saueressig, Taming perturbative divergences in asymptotically safe gravity. Nucl. Phys. B **824**, 168–191 (2010). https://doi.org/10.1016/j.nuclphysb.2009.08.023. arXiv:0902.4630 [hep-th] (cit. on pp. 144, 145)

15. F. Saueressig et al., Higher derivative gravity from the universal renormalization group machine, in *PoS EPS-HEP2011* (2011), p. 124. arXiv:1111.1743 [hep-th] (cit. on pp. 144, 145)
16. D. Benedetti, F. Caravelli, The local potential approximation in quantum gravity. J. High Energy Phys. **6**, 17 (2012). https://doi.org/10.1007/JHEP06(2012)017. arXiv:1204.3541 [hep-th] (cit. on pp. 144, 145)
17. M. Demmel, F. Saueressig, O. Zanusso, Fixed-functionals of three dimensional Quantum Einstein Gravity. J. High Energy Phys. **11**, 131 (2012). https://doi.org/10.1007/JHEP11(2012)131. arXiv:1208.2038 [hep-th] (cit. on pp. 144, 145)
18. J.A. Dietz, T.R. Morris, Asymptotic safety in the f(R) approximation. J. High Energy Phys. **1**, 108 (2013). https://doi.org/10.1007/JHEP01(2013)108. arXiv:1211.0955 [hep-th] (cit. on pp. 144, 145)
19. M. Demmel, F. Saueressig, O. Zanusso, Fixed functionals in asymptotically safe gravity, in *Proceedings of 13th Marcel Grossmann Meeting*, Stockholm, Sweden (2015), pp. 2227–2229. https://doi.org/10.1142/9789814623995_0404. arXiv:1302.1312 [hep-th] (cit. on pp. 144, 145)
20. J.A. Dietz, T.R. Morris, Redundant operators in the exact renormalisation group and in the f(R) approximation to asymptotic safety. J. High Energy Phys. **7**, 64 (2013). https://doi.org/10.1007/JHEP07(2013)064. (cit. on pp. 144, 145)
21. D. Benedetti, F. Guarnieri, Brans-Dicke theory in the local potential approximation. New J. Phys. **16**(5), 053051 (2014). https://doi.org/10.1088/1367-2630/16/5/053051. arXiv:1311.1081 [hep-th] (cit. on pp. 144, 145)
22. M. Demmel, F. Saueressig, O. Zanusso, RG flows of Quantum Einstein Gravity on maximally symmetric spaces. J. High Energy Phys. **6**, 26 (2014). https://doi.org/10.1007/JHEP06(2014)026. arXiv:1401.5495 [hep-th] (cit. on pp. 144, 145)
23. R. Percacci, G.P. Vacca, Search of scaling solutions in scalar-tensor gravity. Eur. Phys. J. C **75**, 188 (2015). https://doi.org/10.1140/epjc/s10052-015-3410-0. arXiv:1501.00888 [hep-th] (cit. on pp. 144, 145)
24. J. Borchardt, B. Knorr, Global solutions of functional fixed point equations via pseudospectral methods. Phys. Rev. D **91**(10), 105011 (2015). https://doi.org/10.1103/PhysRevD.91.105011. arXiv:1502.07511 [hep-th] (cit. on pp. 144, 145)
25. M. Demmel, F. Saueressig, O. Zanusso, A proper fixed functional for four-dimensional Quantum Einstein Gravity. J. High Energy Phys. **8**, 113 (2015). https://doi.org/10.1007/JHEP08(2015)113. arXiv:1504.07656 [hep-th] (cit. on pp. 144, 145)
26. N. Ohta, R. Percacci, G.P. Vacca, Flow equation for f(R) gravity and some of its exact solutions. Phys. Rev. D **92**(6), 061501 (2015). https://doi.org/10.1103/PhysRevD.92.061501. arXiv:1507.00968 [hep-th] (cit. on pp. 144, 145)
27. N. Ohta, R. Percacci, G.P. Vacca, Renormalization group equation and scaling solutions for f(R) gravity in exponential parametrization. Eur. Phys. J. C **76**, 46 (2016). https://doi.org/10.1140/epjc/s10052-016-3895-1. arXiv:1511.09393 [hep-th] (cit. on pp. 144, 145)
28. P. Labus, T.R. Morris, Z.H. Slade, Background independence in a background dependent renormalization group. Phys. Rev. D **94**(2), 024007 (2016). https://doi.org/10.1103/PhysRevD.94.024007. arXiv:1603.04772 [hep-th] (cit. on pp. 144, 145)
29. J.A. Dietz, T.R. Morris, Z.H. Slade, Fixed point structure of the conformal factor field in quantum gravity. Phys. Rev. D **94**(12), 124014 (2016). https://doi.org/10.1103/PhysRevD.94.124014. arXiv:1605.07636 [hep-th] (cit. on pp. 144, 145)
30. E. Manrique, S. Rechenberger, F. Saueressig, Asymptotically safe Lorentzian gravity. Phys. Rev. Lett. **106**(25), 251302 (2011). https://doi.org/10.1103/PhysRevLett.106.251302. arXiv:1102.5012 [hep-th] (cit. on pp. 144-146)
31. S. Rechenberger, F. Saueressig, A functional renormalization group equation for foliated spacetimes. J. High Energy Phys. **3**, 10 (2013). https://doi.org/10.1007/JHEP03(2013)010. arXiv:1212.5114 [hep-th] (cit. on pp. 144, 145)
32. W.B. Houthoff, A. Kurov, F. Saueressig, Impact of topology in foliated Quantum Einstein Gravity. Eur. Phys. J. C **77**, 491 (2017). https://doi.org/10.1140/epjc/s10052-017-5046-8. arXiv:1705.01848 [hep-th] (cit. on pp. 144, 146)

33. P. Donà, A. Eichhorn, R. Percacci, Matter matters in asymptotically safe quantum gravity. Phys. Rev. D **89**(8), 084035 (2014). https://doi.org/10.1103/PhysRevD.89.084035. arXiv:1311.2898 [hep-th] (cit. on p. 145)

34. J. Meibohm, J.M. Pawlowski, M. Reichert, Asymptotic safety of gravity matter systems. Phys. Rev. D **93**(8), 084035 (2016). https://doi.org/10.1103/PhysRevD.93.084035. arXiv:1510.07018 [hep-th] (cit. on p. 145)

35. A. Nink, M. Reuter, The unitary conformal field theory behind 2D Asymptotic Safety. J. High Energy Phys. **2**, 167 (2016). https://doi.org/10.1007/JHEP02(2016)167. arXiv:1512.06805 [hep-th] (cit. on p. 146). [36]

36. H.W. Hamber, R.M. Williams, Simplicial quantum gravity in three dimensions: analytical and numerical results. Phys. Rev. D **47**, 510–532 (1993). https://doi.org/10.1103/PhysRevD.47. 510. (cit. on p. 146)

37. H.W. Hamber, R.M.Williams, Discrete Wheeler-DeWitt equation. Phys. Rev. D **84**(10), 104033 (2011). https://doi.org/10.1103/PhysRevD.84.104033. arXiv:1109.2530 [hep-th] (cit. on p. 146)

38. H.W. Hamber, R. Toriumi, R.M. Williams, Wheeler-DeWitt equation in 2+1 dimensions. Phys. Rev. D **86**(8), 084010 (2012). https://doi.org/10.1103/PhysRevD.86.084010. arXiv:1207.3759 [hep-th] (cit. on p. 146)

39. A. Bonanno, An effective action for asymptotically safe gravity. Phys. Rev. D **85**(8), 081503 (2012). https://doi.org/10.1103/PhysRevD.85.081503. arXiv:1203.1962 [hep-th] (cit. on p. 147)

40. Planck Collaboration et al., Planck 2015 results. XIII. Cosmological parameters. A&A **594**, A13 (2016). https://doi.org/10.1051/0004-6361/201525830. arXiv:1502.01589 (cit. on p. 147)

41. A.A. Starobinsky, A new type of isotropic cosmological models without singularity. Phys. Lett. B **91**, 99–102 (1980). https://doi.org/10.1016/0370-2693(80)90670-X. (cit. on p. 147)

42. A. Eichhorn, A. Held, Top mass from asymptotic safety. Phys. Lett. B **777**, 217–221 (2018). https://doi.org/10.1016/j.physletb.2017.12.040. arXiv:1707.01107 [hep-th] (cit. on p. 147)

43. D. Malafarina, P.S. Joshi, Electromagnetic counterparts to gravitational waves from black hole mergers and naked singularities (2016). arXiv:1603.02848 [gr-qc] (cit. on p. 148)

Appendix A
FRG Equations on Foliated Spacetimes

In order to compute the right-hand-side of the Wetterich equation (4.14) on foliated spacetimes, one needs to compute the second functional derivative $\Gamma_k^{(2)}$ of the action functional (4.21), complemented by appropriate gauge-fixing and ghost contributions. The resulting Hessian must be inverted and the corresponding operator traces have to be evaluate by means of the heat-kernel techniques. This appendix provides all technical details of this calculation.

A.1 Hessians in a FRW Background

The evaluation of the flow equation (4.14) requires the Hessian $\Gamma_k^{(2)}$. The details of this calculation are summarized in this appendix.

In the sequel, indices are raised and lowered with the background metric $\bar{\sigma}_{ij}$. Moreover, we introduce the shorthand notations

$$\int_x \equiv \int d\tau \, d^d y \, \sqrt{\bar{\sigma}} \tag{A.1}$$

and $\hat{\sigma} \equiv \bar{\sigma}^{ij} \hat{\sigma}_{ij}$ to lighten the notation and use $\Delta \equiv -\bar{\sigma}^{ij} \partial_i \partial_j$ to denote the Laplacian on the spatial slices.

A.1.1 Hessians in the Gravitational Sector

When constructing $\Gamma_k^{(2)}$, it is convenient to consider (4.21) as a linear combination of the interaction monomials (4.31). These monomials are then expanded in terms of the fluctuation fields according to

$$N = \bar{N} + \hat{N}, \qquad N_i = \bar{N}_i + \hat{N}_i, \qquad \sigma_{ij} = \bar{\sigma}_{ij} + \hat{\sigma}_{ij}. \tag{A.2}$$

© Springer Nature Switzerland AG 2018

A. B. Platania, *Asymptotically Safe Gravity*, Springer Theses,

https://doi.org/10.1007/978-3-319-98794-1

As an intermediate result, we note that the expansion of the extrinsic curvature (4.12) around the FRW background is given by

$$\delta K_{ij} = -\hat{N}\,\bar{K}_{ij} + \tfrac{1}{2}(\partial_\tau\hat{\sigma}_{ij} - \partial_i\hat{N}_j - \partial_j\hat{N}_i),$$

$$\delta^2 K_{ij} = 2\,\hat{N}^2\,\bar{K}_{ij} - \hat{N}\left(\partial_\tau\hat{\sigma}_{ij} - \partial_i\hat{N}_j - \partial_j\hat{N}_i\right) + \hat{N}^k\left(\partial_i\hat{\sigma}_{jk} + \partial_j\hat{\sigma}_{ik} - \partial_k\hat{\sigma}_{ij}\right),$$

(A.3)

were δ^n denotes the order of the expression in the fluctuation fields. For later reference, it is also useful to have the explicit form of these expressions contracted with the inverse background metric

$$\bar{\sigma}^{ij}\left(\delta K_{ij}\right) = -\hat{N}\bar{K} + \tfrac{1}{2}\bar{\sigma}^{ij}(\partial_\tau\hat{\sigma}_{ij}) - \partial^i\hat{N}_i,$$

$$\bar{\sigma}^{ij}\left(\delta^2 K_{ij}\right) = 2\,\hat{N}^2\bar{K} - \hat{N}\bar{\sigma}^{ij}\left(\partial_\tau\hat{\sigma}_{ij}\right) + 2\hat{N}\partial^i\hat{N}_i + \hat{N}^k\left(2\,\partial^i\hat{\sigma}_{ik} - \partial_k\hat{\sigma}\right).$$

(A.4)

Expanding the interaction monomials (4.31), the terms quadratic in the fluctuation fields are

$$\delta^2 I_1 = \int_x \left[2(\delta K_{ij})\bar{\sigma}^{ik}\bar{\sigma}^{jl}(\delta K_{kl}) + \tfrac{2}{d}\,\bar{K}\,\bar{\sigma}^{ij}\left(\delta^2 K_{ij} + (\delta K_{ij})(2\hat{N} + \hat{\sigma})\right)\right.$$
$$\left. - \tfrac{8}{d}\,\bar{K}\,\hat{\sigma}^{ij}\,(\delta K_{ij}) + \tfrac{1}{d}\,\bar{K}^2\left(\tfrac{d-4}{d}\hat{N}\hat{\sigma} + \tfrac{d-8}{4d}\hat{\sigma}^2 - \tfrac{d-12}{2d}\hat{\sigma}_{ij}\hat{\sigma}^{ij}\right)\right],$$

$$\delta^2 I_2 = \int_x \left[2(\bar{\sigma}^{ij}\delta K_{ij})^2 + 2\bar{K}\,\bar{\sigma}^{ij}\left(\delta^2 K_{ij} + (\delta K_{ij})(2\hat{N} + \hat{\sigma})\right) - 4\bar{K}\hat{\sigma}^{ij}\,(\delta K_{ij})\right.$$
$$\left. + \bar{K}^2\left(\tfrac{d-4}{d}\hat{N}\hat{\sigma} + \tfrac{d^2-8d+8}{4d^2}\hat{\sigma}^2 - \tfrac{d-8}{2d}\hat{\sigma}_{ij}\hat{\sigma}^{ij}\right) - \tfrac{4}{d}\bar{K}\hat{\sigma}\bar{\sigma}^{ij}\delta K_{ij}\right],$$

$$\delta^2 I_3 = \int_x \left[\left(2\hat{N} + \hat{\sigma}\right)\left(\partial_i\partial_j\hat{\sigma}^{ij} + \Delta\hat{\sigma}\right) - \tfrac{1}{2}\hat{\sigma}_{ij}\Delta\hat{\sigma}^{ij} - \tfrac{1}{2}\hat{\sigma}\Delta\hat{\sigma} + \left(\partial_i\hat{\sigma}^{ik}\right)\left(\partial_j\hat{\sigma}^j{}_k\right)\right],$$

$$\delta^2 I_4 = \int_x \left[\hat{N}\hat{\sigma} + \tfrac{1}{4}\hat{\sigma}^2 - \tfrac{1}{2}\hat{\sigma}^{ij}\hat{\sigma}_{ij}\right].$$

(A.5)

In order to arrive at the final form of these expressions, we integrated by parts and made manifest use of the geometric properties of the background (4.24).

In order to develop a consistent gauge-fixing scheme and to simplify the structure of the flow equation it is useful to carry out a further TT decomposition of the fluctuation fields entering into (A.5). A very convenient choice is provided by the standard decomposition of the fluctuation fields used in cosmic perturbation theory where the shift vector and the metric on the spatial slice are rewritten according to (4.34) and (4.35). On this basis the final step expresses the variations (A.5) in terms of the component fields (4.34) and (4.35). The rather lengthy computation can be simplified by using the identities

$$\hat{\sigma} = -(d-1)\psi - E, \qquad \partial^i\hat{\sigma}_{ij} = -\partial_j E - \sqrt{\Delta}\,v_j, \qquad \partial^i\partial^j\hat{\sigma}_{ij} = \Delta E, \quad \text{(A.6)}$$

together with the relations (4.37) and (A.6). Starting with the kinetic terms appearing in $\delta^2 I_1$ and $\delta^2 I_2$,

$$\mathbb{K}_1 \equiv 2 \int_x (\delta K_{ij}) \, \bar{\sigma}^{ik} \bar{\sigma}^{jl} \, (\delta K_{kl}), \qquad \mathbb{K}_2 \equiv 2 \int_x (\bar{\sigma}^{ij} \, \delta K_{ij})^2, \qquad \text{(A.7)}$$

the resulting expressions written in terms of component fields are

$$
\begin{aligned}
\mathbb{K}_1 = \int_x \Big[&-\tfrac{1}{2} h^{ij} \left(\partial_\tau^2 + \tfrac{d-4}{d} \bar{K} \partial_\tau \right) h_{ij} - \tfrac{d-1}{2} \psi \left(\partial_\tau + \tfrac{d-2}{d} \bar{K}\right) \left(\partial_\tau + \tfrac{2}{d} \bar{K}\right) \psi \\
&-\tfrac{1}{2} E \left(\partial_\tau + \tfrac{d-2}{d} \bar{K}\right) \left(\partial_\tau + \tfrac{2}{d} \bar{K}\right) E - v^i \left(\partial_\tau + \tfrac{d-3}{d} \bar{K}\right) \left(\partial_\tau + \tfrac{1}{d} \bar{K}\right) v_i \\
&- 2 B \sqrt{\Delta} \left(\partial_\tau + \tfrac{2}{d} \bar{K}\right) E - 2 u^i \left(\partial_\tau + \tfrac{2}{d} \bar{K}\right) \sqrt{\Delta} \, v_i + 2 B \Delta B \\
&+ u_i \Delta u^i - \tfrac{4}{d} \bar{K} \hat{N} \sqrt{\Delta} B + \tfrac{2}{d} \bar{K} \hat{N} \left(\partial_\tau + \tfrac{2}{d} \bar{K}\right) \big((d-1)\psi + E\big) \\
&+ \tfrac{2}{d} \bar{K}^2 \hat{N}^2 \Big],
\end{aligned}
$$

$$\text{(A.8)}$$

and

$$
\begin{aligned}
\mathbb{K}_2 = \int_x \Big[&-\tfrac{1}{2}\big((d-1)\psi + E\big) \left(\partial_\tau + \tfrac{d-2}{d} \bar{K}\right) \left(\partial_\tau + \tfrac{2}{d} \bar{K}\right) \big((d-1)\psi + E\big) \\
&- 2 B \sqrt{\Delta} \left(\partial_\tau + \tfrac{2}{d} \bar{K}\right) \big((d-1)\psi + E\big) + 2 B \Delta B \\
&- 4 \bar{K} \hat{N} \sqrt{\Delta} B + 2 \bar{K} \hat{N} \left(\partial_\tau + \tfrac{2}{d} \bar{K}\right) \big((d-1)\psi + E\big) + 2 \bar{K}^2 \hat{N}^2 \Big].
\end{aligned}
$$

$$\text{(A.9)}$$

On this basis one finds that

$$
\begin{aligned}
\delta^2 I_1 = \mathbb{K}_1 - \int_x \Big[&+ \tfrac{(d-4)(d-8)}{4d^2} \bar{K}^2 \hat{N} \big((d-1)\psi + E\big) \\
&+ \tfrac{4}{d} \bar{K} h^{ij} (\partial_\tau + \tfrac{d-12}{8d} \bar{K}) h_{ij} \\
&- \tfrac{4}{d} \bar{K} \big(u^k \sqrt{\Delta} v_k + B \sqrt{\Delta} E - v^i (\partial_\tau + \tfrac{d-8}{4d} \bar{K}) v_i\big) \\
&- \tfrac{1}{d} \bar{K} \big((d-1)\psi + E\big)(\partial_\tau + \tfrac{1}{4}\bar{K})\big((d-1)\psi + E\big) \\
&+ \tfrac{4}{d} \bar{K} E (\partial_\tau + \tfrac{d+4}{8d} \bar{K}) E + \tfrac{4(d-1)}{d} \bar{K} \psi (\partial_\tau + \tfrac{d+4}{8d} \bar{K}) \psi \Big]
\end{aligned}
$$

$$\text{(A.10)}$$

and

$$
\begin{aligned}
\delta^2 I_2 = \mathbb{K}_2 - \int_x \Big[&- \tfrac{d-4}{d} \bar{K}^2 \hat{N}\big((d-1)\psi + E\big) + 2 \bar{K} h^{ij} (\partial_\tau + \tfrac{d-8}{4d} \bar{K}) h_{ij} \\
&- \bar{K} \big((d-1)\psi + E\big)(\tfrac{d-2}{d} \partial_\tau + \tfrac{d^2-8}{4d^2} \bar{K})\big((d-1)\psi + E\big) \\
&+ 2 \bar{K} E (\partial_\tau + \tfrac{1}{4} \bar{K}) E + 2(d-1)\bar{K} \psi (\partial_\tau + \tfrac{1}{4} \bar{K}) \psi \\
&+ 2 \bar{K} v^i (\partial_\tau + \tfrac{d-6}{2d} \bar{K}) v_i - \tfrac{4}{d} \bar{K} \big((d-1)\psi + E\big) \sqrt{\Delta} B \Big].
\end{aligned}
$$

$$\text{(A.11)}$$

Finally, $\delta^2 I_3$ and $\delta^2 I_4$ written in terms of the component fields are

$$
\begin{aligned}
\delta^2 I_3 &= \int_x \Big[\tfrac{(d-1)(d-2)}{2} \psi \Delta \psi - \tfrac{1}{2} h_{ij} \Delta h^{ij} - 2(d-1) \hat{N} \Delta \psi \Big], \\
\delta^2 I_4 &= \int_x \Big[\tfrac{(d-1)(d-3)}{4} \psi^2 + \tfrac{(d-1)}{2} \psi E - \tfrac{1}{4} E^2 - \hat{N}\big((d-1)\psi + E\big) - \tfrac{1}{2} h_{ij} h^{ij} - v_i v^i \Big].
\end{aligned}
$$

$$\text{(A.12)}$$

Combining these variations according to (4.21) one arrives the matrix entries for $\delta^2 \Gamma_k^{\text{grav}}$. In the flat space limit, where $\bar{K} = 0$, these entries are listed in the second column of Table 4.2.

A.1.2 Gauge-Fixing Terms

Following the strategy of the previous subsection it is useful to also decompose the gauge-fixing terms (4.38) into two interaction monomials

$$\delta^2 I_5 \equiv \int_x F^2\,,$$
$$\delta^2 I_6 \equiv \int_x F_i\,\bar{\sigma}^{ij}\,F_j\,. \tag{A.13}$$

Since the functionals F and F_i defined in Eq. (4.39) are linear in the fluctuation fields, the gauge-fixing terms are quadratic in the fluctuations by construction. This feature is highlighted by adding the δ^2 to the definition of the monomials.

Substituting the explicit form of F and F_i and recasting the resulting expressions in terms of the component fields (4.34) and (4.35) one finds

$$\delta^2 I_5 = \int_x \Big[c_2^2\, B\Delta B - \hat{N}\,\big(c_1\partial_\tau + (c_1 - c_9)\bar{K}\big)\big(c_1\partial_\tau + c_9\bar{K}\big)\hat{N}$$
$$- \big((d-1)\psi + E\big)\big(c_3\partial_\tau + (c_3 - c_8)\bar{K}\big)\big(c_3\partial_\tau + c_8\bar{K}\big)\big((d-1)\psi + E\big)$$
$$- 2c_2\,B\sqrt{\Delta}\,\big(c_1\,\partial_\tau + c_9\bar{K}\big)\hat{N}$$
$$+ 2\hat{N}\,\big(c_1\partial_\tau + (c_1 - c_9)\bar{K}\big)\big(c_3\partial_\tau + c_8\bar{K}\big)\big((d-1)\psi + E\big)$$
$$+ 2c_2\,B\sqrt{\Delta}\,\big(c_3\partial_\tau + c_8\bar{K}\big)\big((d-1)\psi + E\big)\Big] \tag{A.14}$$

and

$$\delta^2 I_6 = \int_x \Big[c_5^2\,\hat{N}\Delta\hat{N} - u^i\,\big(c_4\,\partial_\tau + (\tfrac{d-2}{d}c_4 - c_{10})\,\bar{K}\big)\big(c_4\partial_\tau + c_{10}\bar{K}\big)u_i$$
$$- B\big(c_4\,\partial_\tau + (\tfrac{d-1}{d}c_4 - c_{10})\,\bar{K}\big)\big(c_4\partial_\tau + (\tfrac{1}{d}c_4 + c_{10})\,\bar{K}\big)B$$
$$+ 2c_5\,\hat{N}\big(c_4\partial_\tau + (\tfrac{2}{d}c_4 + c_{10})\,\bar{K}\big)\sqrt{\Delta}B$$
$$- 2c_6\,\big((d-1)\psi + E\big)\big(c_4\partial_\tau + (\tfrac{2}{d}c_4 + c_{10})\bar{K}\big)\sqrt{\Delta}B$$
$$- 2c_7\,E\big(c_4\partial_\tau + (\tfrac{2}{d}c_4 + c_{10})\bar{K}\big)\sqrt{\Delta}B$$
$$- 2c_7\,v^i\,\sqrt{\Delta}\,\big(c_4\partial_\tau + c_{10}\bar{K}\big)u_i - 2c_5c_6\,\hat{N}\Delta\big((d-1)\psi + E\big)$$
$$- 2c_5\,c_7\,\hat{N}\Delta E + c_6^2\,\big((d-1)\psi + E\big)\Delta\big((d-1)\psi + E\big)$$
$$+ 2c_6c_7\,\big((d-1)\psi + E\big)\Delta E + c_7^2\,\big(E\Delta E + v^i\Delta v_i\big)\Big]. \tag{A.15}$$

Here again we made use of the geometric properties of the background and integrated by parts in order to obtain a similar structure as in the gravitational sector.

Combining the results (A.10), (A.11), (A.12), (A.14), and (A.15), taking into account the relative signs between the terms and restoring the coupling constants according to (4.33) gives the part of the gauge-fixed gravitational action quadratic in the fluctuation fields. The explicit result is rather lengthy and given by

$$
\begin{aligned}
32\pi G_k &\left(\tfrac{1}{2}\delta^2\Gamma_k^{\mathrm{grav}} + \Gamma_k^{\mathrm{gf}}\right) = \\
\int_x &\Bigg\{ -\hat{N}\Big[(c_1\partial_\tau + (c_1 - c_9)\bar{K})(c_1\partial_\tau + c_9\bar{K}) - c_5^2\Delta + \tfrac{2(d-1)}{d}\bar{K}^2\Big]\hat{N} \\
&- B\Big[(c_4\partial_\tau + (\tfrac{d-1}{d}c_4 - c_{10})\bar{K})(c_4\partial_\tau + (\tfrac{1}{d}c_4 + c_{10})\bar{K}) - c_2^2\Delta\Big]B \\
&- 2B\sqrt{\Delta}\Big[(c_1c_2 + c_4c_5)\partial_\tau + (c_2c_9 + c_4c_5\tfrac{d-2}{d} - c_5c_{10} - \tfrac{2(d-1)}{d})\bar{K}\Big]\hat{N} \\
&+ 2\hat{N}\Big[(c_1\partial_\tau + (c_1 - c_9)\bar{K})(c_3\partial_\tau + c_8\bar{K}) - \tfrac{d-1}{d}\bar{K}\partial_\tau - c_5(c_6 + c_7)\Delta \\
&\qquad - \tfrac{5d^2-12d+16}{8d^2}\bar{K}^2 - \Lambda_k\Big]E \\
&+ 2(d-1)\hat{N}\Big[(c_1\partial_\tau + (c_1 - c_9)\bar{K})(c_3\partial_\tau + c_8\bar{K}) - \tfrac{d-1}{d}\bar{K}\partial_\tau \\
&\qquad + (1 - c_5c_6)\Delta - \tfrac{5d^2-12d+16}{8d^2}\bar{K}^2 - \Lambda_k\Big]\psi \\
&+ 2B\sqrt{\Delta}\Big[(c_2c_3 + c_4(c_6 + c_7))\partial_\tau + (c_2c_8 + (c_6 + c_7)(\tfrac{d-2}{d}c_4 - c_{10}))\bar{K}\Big]E \\
&+ 2(d-1)B\sqrt{\Delta}\Big[(1 + c_2c_3 + c_4c_6)\partial_\tau + (c_2c_8 + \tfrac{d-2}{d}c_4c_6 - c_6c_{10})\bar{K}\Big]\psi \\
&- (d-1)\psi\Big[(d-1)\big((c_3\partial_\tau + (c_3 - c_8)\bar{K})(c_3\partial_\tau + c_8\bar{K}) - c_6^2\Delta\big) \\
&\qquad + \tfrac{d-2}{2}(-\partial_\tau^2 + \Delta - \tfrac{2}{d}\dot{\bar{K}}) + \tfrac{d^2-10d+14}{2d}\bar{K}\partial_\tau + \tfrac{d^2-8d+11}{4d}\bar{K}^2 - \tfrac{d-3}{2}\Lambda_k\Big]\psi \\
&+ E\Big[(c_6 + c_7)^2\Delta - (c_3\partial_\tau + (c_3 - c_8)\bar{K})(c_3\partial_\tau + c_8\bar{K}) \\
&\qquad - \tfrac{1}{2}\Lambda_k + \tfrac{d-1}{d}\bar{K}\partial_\tau + \tfrac{d-1}{4d}\bar{K}^2\Big]E \\
&+ (d-1)\psi\Big[2c_6(c_6 + c_7)\Delta - 2(c_3\partial_\tau + (c_3 - c_8)\bar{K})(c_3\partial_\tau + c_8\bar{K}) \\
&\qquad + \partial_\tau^2 + \bar{K}\partial_\tau + \tfrac{d-1}{d}\dot{\bar{K}} + \tfrac{d-1}{2d}\bar{K}^2 + \Lambda_k\Big]E \\
&- u^i\Big[(c_4\partial_\tau + (\tfrac{d-2}{d}c_4 - c_{10})\bar{K})(c_4\partial_\tau + c_{10}\bar{K}) - \Delta\Big]u_i \\
&+ v^i\Big[-\partial_\tau^2 + \tfrac{d-2}{d}\bar{K}\partial_\tau - \tfrac{1}{d}\dot{\bar{K}} + \tfrac{d^2-8d+11}{d^2}\bar{K}^2 + c_7^2\Delta - 2\Lambda_k\Big]v_i \\
&- 2u^i\Big[(1 - c_4c_7)\partial_\tau + c_7(c_{10} - \tfrac{d-2}{d}c_4)\bar{K}\Big]\sqrt{\Delta}\,v_i \\
&+ \tfrac{1}{2}h^{ij}\Big[-\partial_\tau^2 + \tfrac{3d-4}{d}\bar{K}\partial_\tau + \tfrac{d^2-9d+12}{d^2}\bar{K}^2 + \Delta - 2\Lambda_k\Big]h_{ij}\Bigg\}.
\end{aligned}
$$

$$\tag{A.16}$$

Based on this general result, one may then search for a particular gauge fixing which, firstly, eliminates all terms containing $\sqrt{\Delta}$ and, secondly, ensures that all component fields obey a relativistic dispersion relation in the limit when $\bar{K} = 0$. A careful inspection of Eq. (A.16) shows that there is an essentially unique gauge choice which satisfies both conditions. The resulting values for the coefficients c_i are given in Eq. (4.40). Specifying the general result to these values finally results in the gauge-fixed Hessian appearing in the gravitational sector (4.44). Taking the limit $\bar{K} = 0$, the propagators resulting from this gauge-fixing are displayed in the third column of Table 4.2. In this way it is straightforward to verify that the gauge choice indeed satisfies the condition of a relativistic dispersion relation for all component fields.

The gauge-fixing is naturally accompanied by a ghost action exponentiating the resulting Faddeev-Popov determinant. For the gauge-fixing conditions F and F_i the ghost action comprises a scalar ghost $\{\bar{c}, c\}$ and a (spatial) vector ghost $\{\bar{b}_i, b_i\}$. Their action can be constructed in a standard way by evaluating

$$\Gamma_k^{\text{scalar ghost}} = \int_x \bar{c}\, \frac{\delta F}{\delta \hat{\chi}^i}\, \delta_{c, b_i} \chi^i\,, \qquad \Gamma_k^{\text{vector ghost}} = \int_x \bar{b}^j\, \frac{\delta F_j}{\delta \hat{\chi}^i}\, \delta_{c, b_i} \chi^i\,. \qquad (A.17)$$

Here $\frac{\delta F}{\delta \hat{\chi}^i}$ denotes the variation of the gauge-fixing condition with respect to the fluctuation fields $\hat{\chi} = \left\{\hat{N}, \hat{N}_i, \hat{\sigma}_{ij}\right\}$ at fixed background and the expressions $\delta_{c, b_i} \chi^i$ are given by the variations (4.8) with the parameters f and ζ_i replaced by the scalar ghost c and vector ghost b_i, respectively. Taking into account terms quadratic in the fluctuation fields only, the resulting ghost action is given in Eq. (4.45). Together with the Hessian in the gravitational sector, Eq. (4.44), this result completes the construction of the Hessians entering the right-hand-side of the flow Eq. (4.14).

A.2   Evaluation of the Operator Traces

In Sect. 4.2 the operator traces have been written in terms of the standard $D = (d + 1)$-dimensional Laplacian $\Box_s \equiv -\bar{g}^{\mu\nu} D_\mu D_\nu$ where $s = 0, 1, 2$ indicates that the Laplacian is acting on fields with zero, one or two spatial indices. In this appendix, we use the heat-kernel techniques introduced in Sect. 3.2.3 to construct the resulting contributions to the flow.

A.2.1   Cutoff Scheme and Master Traces

The final step in the construction of the right-hand-side of the flow equation is the specification of the regulator $\mathcal{R}_k$. In particular we will resort to regulators of Type I, which are implicitly defined through the relation that the regulator dresses up each D-dimensional Laplacian by a scale-dependent mass term according to the rule

$$\Box_s \mapsto P_k \equiv \Box_s + R_k \,. \tag{A.18}$$

The prescription (A.18) then fixes the matrix-valued regulator $\mathcal{R}_k$ uniquely. Following the strategy of Sect. 3.2.3, the resulting modified inverse propagator can be decomposed as $\widetilde{\Gamma}_k^{(2)} = \hat{L}_k(\widetilde{\mathcal{P}} + \widetilde{\mathcal{V}})$. In the case at hand the interaction matrix $\widetilde{\mathcal{V}}$ is a function of the extrinsic background curvature $\bar{K}$, while $\widetilde{\mathcal{P}}$ has the form (3.47). Accordingly, the matrix elements of $\Gamma_k^{(2)}$ found in Sect. A.1 can be cast in the form

$$\Gamma_k^{(2)}\Big|_{ij} = (32\pi G_k)^{-\alpha_i} \, c_i \left[\Box_{s_i} + w + \widetilde{\mathcal{V}}_{ij}(\bar{K}) \right] \,, \tag{A.19}$$

where α_i and w are the same parameters appearing in Eqs. (3.46) and (3.47), while the interaction operator $\widetilde{\mathcal{V}}(\bar{K})$ has the following structure

$$\widetilde{\mathcal{V}}(\bar{K}) = \hat{v}_1 \, \bar{K}^2 + \hat{v}_2 \, \dot{\bar{K}} + \hat{v}_3 \, \bar{K} \partial_\tau \,. \tag{A.20}$$

Notably, c_i and $(\hat{v}_l)_{ij}$ are d-dependent numerical coefficients whose values can be read off from Eqs. (4.44) and (4.45). Applying the rule (A.18) then yields

$$\mathcal{R}_k|_{ij} = (32\pi G_k)^{-\alpha_i} \, c_i \, R_k \,. \tag{A.21}$$

Subsequently, one has to construct the inverse of $\widetilde{\Gamma}_k^{(2)}$. Given the left-hand-side of the flow Eq. (4.25) it thereby suffices to keep track of terms containing up to two time-derivatives of the background quantities, i.e., $\bar{K}^2$ and $\dot{\bar{K}}$. Hence, defining $\mathcal{P} = \hat{L}_k \widetilde{\mathcal{P}}$ and $\mathcal{V} = \hat{L}_k \widetilde{\mathcal{V}}$, the effective inverse propagator is conveniently written as

$$\widetilde{\Gamma}_k^{(2)} \equiv \left(\Gamma_k^{(2)} + \mathcal{R}_k \right) \equiv \mathcal{P} + \mathcal{V} \,, \tag{A.22}$$

where the matrix $\mathcal{P}$ collects all terms containing $\Box_{s_i}$ and Λ_k and the potential-matrix $\mathcal{V}$ collects the terms with at least one power of the extrinsic background curvature $\bar{K}$. The inverse of the modified Hessian $\widetilde{\Gamma}_k^{(2)}$ can then be constructed as an expansion in $\mathcal{V}$. Retaining terms containing up to two powers of $\bar{K}$ only

$$\left(\mathcal{P} + \mathcal{V} \right)^{-1} = \mathcal{P}^{-1} - \mathcal{P}^{-1} \mathcal{V} \mathcal{P}^{-1} + \mathcal{P}^{-1} \mathcal{V} \mathcal{P}^{-1} \mathcal{V} \mathcal{P}^{-1} + O(\bar{K}^3) \,. \tag{A.23}$$

At this stage it is instructive to look at a *single block* for which we assume that it is spanned by a single field $\hat{\chi}_i$ having s indices. In a slight abuse of notation we denote the kinetic and potential operators on this block by $\mathcal{P}$ and $\mathcal{V}$. From the structure of the Hessians one finds that the propagator for a single field has the form

$$\mathcal{P}^{-1} = (32\pi G_k)^{\alpha_s} c^{-1} \left(\Box_s + R_k + w \right)^{-1} \,, \tag{A.24}$$

while the potential $\mathcal{V}$ is constructed from three different types of insertions

$$\mathcal{V}_1 = (32\pi G_k)^{-\alpha_s} c \, \bar{K}^2 \, , \quad \mathcal{V}_2 = (32\pi G_k)^{-\alpha_s} c \, \dot{\bar{K}} \, , \quad \mathcal{V}_3 = (32\pi G_k)^{-\alpha_s} c \, \bar{K} \partial_\tau \, .$$
(A.25)

The structure (A.23) can be used to write the right-hand-side of the flow equation in terms of master traces, which are independent of the particular choice of cutoff function. Notably, it is convenient to express the operator traces in terms of the threshold functions defined in Eq. (3.48). In particular, for a cutoff of Litim type

$$R_k = (k^2 - \Box_s) \, \theta(k^2 - \Box_s)$$
(A.26)

to which we resort throughout this thesis, the integrals in the threshold functions can be evaluated analytically and yield

$$\Phi_n^p(w) \equiv \frac{1}{\Gamma(n+1)} \frac{1}{(1+w)^p} \, , \quad \widetilde{\Phi}_n^p(w) \equiv \frac{1}{\Gamma(n+2)} \frac{1}{(1+w)^p} \, .$$
(A.27)

The right-hand-side of the flow equation is then conveniently evaluated in terms of the following master traces. For zero potential insertions one has

$$\mathrm{Tr}\left[\mathcal{P}^{-1} \partial_t R_k\right] = \frac{k^D}{(4\pi)^{D/2}} \int_x \Big[+ a_0 \left(2\Phi_{D/2}^1(\tilde{w}) - \eta\,\alpha_s\,\widetilde{\Phi}_{D/2}^1(\tilde{w})\right)$$
$$+ a_2 \left(2\Phi_{D/2-1}^1(\tilde{w}) - \eta\,\alpha_s\,\widetilde{\Phi}_{D/2-1}^1(\tilde{w})\right) \frac{\bar{K}^2}{k^2} \Big] \, .$$
(A.28)

The case with one potential insertion gives

$$\mathrm{Tr}\left[\mathcal{P}^{-1} \mathcal{V}_1 \mathcal{P}^{-1} \partial_t R_k\right] = + \frac{k^D}{(4\pi)^{D/2}} \int_x a_0 \left(2\Phi_{D/2}^2(\tilde{w}) - \eta\,\alpha_s\,\widetilde{\Phi}_{D/2}^2(\tilde{w})\right) \frac{\bar{K}^2}{k^2} \, ,$$

$$\mathrm{Tr}\left[\mathcal{P}^{-1} \mathcal{V}_2 \mathcal{P}^{-1} \partial_t R_k\right] = - \frac{k^D}{(4\pi)^{D/2}} \int_x a_0 \left(2\Phi_{D/2}^2(\tilde{w}) - \eta\,\alpha_s\,\widetilde{\Phi}_{D/2}^2(\tilde{w})\right) \frac{\bar{K}^2}{k^2} \, ,$$

$$\mathrm{Tr}\left[\mathcal{P}^{-1} \mathcal{V}_3 \mathcal{P}^{-1} \partial_t R_k\right] = 0 \, .$$
(A.29)

At the level of two insertions only the trace containing $(\mathcal{V}_3)^2$ contributes to the flow. In this case, the application of off-diagonal heat-kernel techniques yields

$$\mathrm{Tr}\left[(\mathcal{V}_3)^2 \mathcal{P}^{-3} \partial_t R_k\right] = -\frac{1}{2} \frac{k^D}{(4\pi)^{D/2}} \int_x a_0 \left(2\Phi_{D/2+1}^2(\tilde{w}) - \eta\,\alpha_s\,\widetilde{\Phi}_{D/2+1}^2(\tilde{w})\right) \frac{\bar{K}^2}{k^2} \, .$$
(A.30)

Here a_0 and a_2 are the spin-dependent heat-kernel coefficients introduced in Sect. 4.2.2 and $\tilde{w} \equiv w k^{-2}$. Note that once a trace contains two derivatives of the background curvature, all remaining derivatives may be commuted freely, since commutators give rise to terms which do not contribute to the flow of G_k and Λ_k.

A.2.2 Trace Contributions in the Gravitational Sector

At this stage, we have all the ingredients for evaluating the operator traces appearing on the right-hand-side of the FRGE, keeping all terms contributing to the truncation (4.47). In order to cast the resulting expressions into compact form, it is convenient to combine the threshold functions (A.27) according to

$$q_n^p(w) \equiv 2\, \Phi_n^p(w) - \eta\, \tilde{\Phi}_n^p(w)\,, \tag{A.31}$$

and recall the definition of the dimensionless quantities (4.47). Moreover, all traces include the proper factors of $1/2$ and signs appearing on the right-hand-side of the FRGE.

We first evaluate the traces arising from the blocks of $(\Gamma^{(2)} + \mathcal{R}_k)$ which are one-dimensional in field space. In the gravitational sector, this comprises the contributions of the component fields h_{ij}, u_i, v_i, and B. Applying the master formulas (A.28) and (A.29) and adding the results, one has

$$\mathrm{Tr}|_{hh} = \frac{k^D}{2\,(4\pi)^{D/2}} \int_x \left[\frac{(d+1)(d-2)}{2}\, q_{D/2}^1(-2\lambda) + \frac{d^4-2d^3-d^2+14d+36}{12d^2}\, q_{D/2-1}^1(-2\lambda)\, \frac{\bar{K}^2}{k^2} \right.$$
$$\left. - \frac{(d-2)^2(d+1)^2}{2d^2}\, q_{D/2}^2(-2\lambda)\, \frac{\bar{K}^2}{k^2} \right],$$

$$\mathrm{Tr}|_{vv} = \frac{k^D}{2\,(4\pi)^{D/2}} \int_x \left[(d-1)\, q_{D/2}^1(-2\lambda) + \frac{d^3-2d^2+d+6}{6d^2}\, q_{D/2-1}^1(-2\lambda)\, \frac{\bar{K}^2}{k^2} \right.$$
$$\left. - \frac{(d-1)(d^2-5d+7)}{d^2}\, q_{D/2}^2(-2\lambda)\, \frac{\bar{K}^2}{k^2} \right],$$

$$\mathrm{Tr}|_{uu} = \frac{k^D}{2\,(4\pi)^{D/2}} \int_x \left[(d-1)\, q_{D/2}^1(0) + \frac{d^3-2d^2+d+6}{6d^2}\, q_{D/2-1}^1(0)\, \frac{\bar{K}^2}{k^2} \right.$$
$$\left. - \frac{(d-1)(d-2)}{d}\, q_{D/2}^2(0)\, \frac{\bar{K}^2}{k^2} \right],$$

$$\mathrm{Tr}|_{BB} = \frac{k^D}{2\,(4\pi)^{D/2}} \int_x \left[q_{D/2}^1(0) + \frac{d-1}{6d}\, q_{D/2-1}^1(0)\, \frac{\bar{K}^2}{k^2} - \frac{(d-1)^2}{d^2}\, q_{D/2}^2(0)\, \frac{\bar{K}^2}{k^2} \right]. \tag{A.32}$$

The evaluation of the traces in the ghost sector follows along the same lines. In this case one also has a contribution from the third master trace (A.30). The total contributions of the scalar ghosts is then given by

$$-\mathrm{Tr}|_{\bar{c}c} = -\frac{k^D}{(4\pi)^{D/2}} \int_x \left\{ 2\, \Phi_{D/2}^1 + \frac{\bar{K}^2}{k^2} \left[\frac{d-1}{3d}\, \Phi_{D/2-1}^1 + 2\Phi_{D/2}^1 - \frac{4}{d^2}\, \Phi_{D/2+1}^1 \right] \right\}, \tag{A.33}$$

where all threshold functions are evaluated at zero argument. Recalling that the vector ghost b_i is not subject to a transverse constraint, the trace evaluates to

$$-\mathrm{Tr}|_{\bar{b}b} = -\frac{k^D}{(4\pi)^{D/2}} \int_x \left\{ 2d\, \Phi_{D/2}^1 + \frac{\bar{K}^2}{k^2} \left[\frac{d-1}{3}\, \Phi_{D/2-1}^1 + \frac{8}{d}\, \Phi_{D/2}^1 - \frac{4}{d}\, \Phi_{D/2+1}^1 \right] \right\}. \tag{A.34}$$

The last contribution of the flow is provided by the three scalar fields $\xi = (\hat{N}, E, \psi)$. Inspecting (4.44), one finds that the block $(\Gamma^{(2)} + \mathcal{R}_k)$ appearing in this sector is given by a (3×3)-matrix in field space with non-zero off-diagonal entries. Applying the decomposition (A.22) the matrix $\mathcal{P}$ resulting from (4.44) is

$$\mathcal{P} = (32\pi G_k)^{-1} \begin{bmatrix} \Box_0 & \frac{1}{2}(\Box_0 - 2\Lambda) & \frac{d-1}{2}(\Box_0 - 2\Lambda) \\ \frac{1}{2}(\Box_0 - 2\Lambda) & \frac{1}{4}(\Box_0 - 2\Lambda) & -\frac{d-1}{4}(\Box_0 - 2\Lambda) \\ \frac{d-1}{2}(\Box_0 - 2\Lambda) & -\frac{d-1}{4}(\Box_0 - 2\Lambda) & -\frac{(d-1)(d-3)}{4}(\Box_0 - 2\Lambda) \end{bmatrix},$$

(A.35)

while the matrix $\mathcal{V}$ is symmetric with entries

$$\begin{aligned} \mathcal{V}_{11} &= -\frac{2(d-1)}{d^2}\left(2\bar{K}^2 + d\dot{\bar{K}}\right), & \mathcal{V}_{12} &= -\frac{5d^2 - 12d + 16}{8d^2}\bar{K}^2 \\ \mathcal{V}_{22} &= -\frac{d-1}{4d}\left(\bar{K}^2 + 2\dot{\bar{K}}\right), & \mathcal{V}_{13} &= -\frac{(d-1)(5d^2 - 12d + 16)}{8d^2}\bar{K}^2 \\ \mathcal{V}_{33} &= \frac{(d-3)(d-1)^2}{4d}\left(\bar{K}^2 + 2\dot{\bar{K}}\right), & \mathcal{V}_{23} &= \frac{(d-1)^2}{4d}\left(\bar{K}^2 + 2\dot{\bar{K}}\right). \end{aligned}$$

(A.36)

Applying (A.18), the cutoff $\mathcal{R}_k$ in this sector is given by

$$\mathcal{R}_k = (32\pi G_k)^{-1} R_k \begin{bmatrix} 1 & \frac{1}{2} & \frac{d-1}{2} \\ \frac{1}{2} & \frac{1}{4} & -\frac{d-1}{4} \\ \frac{d-1}{2} & -\frac{d-1}{4} & -\frac{(d-1)(d-3)}{4} \end{bmatrix}.$$

(A.37)

The master traces (A.28) and (A.29) also hold in the case where $\mathcal{P}$ and $\mathcal{V}$ are matrix valued. Constructing the inverse of $\mathcal{P}$ on field space explicitly and evaluating the corresponding traces, the contribution of this block to the flow is found as

$$\begin{aligned} \mathrm{Tr}|_{\xi\xi} = \frac{k^D}{2(4\pi)^{D/2}} \int_x \Big[&2q^1_{D/2}(-2\lambda) + q^1_{D/2}\left(-\frac{d}{d-1}\lambda\right) \\ &+ \frac{d-1}{6d}\left(2q^1_{D/2-1}(-2\lambda) + q^1_{D/2-1}\left(-\frac{d}{d-1}\lambda\right)\right)\frac{\bar{K}^2}{k^2} \\ &- \left(\frac{2(d-1)}{d}q^2_{D/2}(-2\lambda) - \frac{3d^3 + 6d^2 - 16d + 16}{4d^2(d-1)}q^2_{D/2}\left(-\frac{d}{d-1}\lambda\right)\right)\frac{\bar{K}^2}{k^2}\Big]. \end{aligned}$$

(A.38)

The traces (A.32), (A.33), (A.34), and (A.38) complete the evaluation of the flow equation on a flat FRW background. Substituting these expressions into the FRGE (4.14) and retaining the terms present in (4.25) then leads to the beta functions (4.49) where the threshold functions are evaluated with a Litim type regulator (A.27).

A.2.3 Minimally Coupled Matter Fields

At the level of the Einstein-Hilbert truncation (4.21), including the contribution of the matter sector (4.18) to the flow of Newton's constant and the cosmological constant is rather straightforward. When expanding the matter fields around a vanishing background value, the Hessian $\Gamma_k^{(2)}$ arising in the matter sector contains variations with respect to the matter fields only and all Laplacians reduce to the background Laplacians. The resulting contributions of the matter trace are then identical to the ones obtained in the metric formulation [1–3]. The trace capturing the contributions of the N_S scalar fields ϕ yields

$$\mathrm{Tr}|_{\phi\phi} = N_S \frac{k^D}{(4\pi)^{D/2}} \int_x \left\{ \Phi_{D/2}^1(0) + \frac{d-1}{6d} \, \Phi_{D/2-1}^1(0) \, \frac{\bar{K}^2}{k^2} \right\}. \tag{A.39}$$

The gauge sector, comprising N_V gauge fields A_μ and the corresponding Faddeev-Popov ghosts $\{\bar{C}_\mu, C_\mu\}$ contributes

$$\mathrm{Tr}|_{AA} = N_V \frac{k^D}{(4\pi)^{D/2}} \int_x \left\{ (d+1) \, \Phi_{D/2}^1(0) + \frac{(d-1)(d^2+2d-11)}{6d\,(d+1)} \, \Phi_{D/2-1}^1(0) \, \frac{\bar{K}^2}{k^2} \right\}, \tag{A.40}$$

and

$$-\mathrm{Tr}|_{\bar{C}C} = N_V \frac{k^D}{(4\pi)^{D/2}} \int_x \left\{ 2 \, \Phi_{D/2}^1(0) + \frac{d-1}{3d} \, \Phi_{D/2-1}^1(0) \, \frac{\bar{K}^2}{k^2} \right\}. \tag{A.41}$$

Adding Eqs. (A.40) and (A.41) gives the total contribution of the gauge fields to the renormalization group flow

$$\mathrm{Tr}|_{\mathrm{GF}} = N_V \frac{k^D}{(4\pi)^{D/2}} \int_x \left\{ (d-1) \, \Phi_{D/2}^1(0) + \frac{(d-1)(d^2-13)}{6d\,(d+1)} \, \Phi_{D/2-1}^1(0) \, \frac{\bar{K}^2}{k^2} \right\}. \tag{A.42}$$

When evaluating the contribution of the fermionic degrees of freedom, we follow the discussion [3], resulting in

$$\mathrm{Tr}|_{\psi\psi} = -\frac{N_D \, 2^{(d+1)/2} \, k^D}{(4\pi)^{D/2}} \int_x \left\{ \Phi_{D/2}^1(0) + \frac{d-1}{d}\left[\left(\tfrac{1}{6} - \tfrac{r}{4}\right) \Phi_{D/2-1}^1(0) - \tfrac{1-r}{4}\Phi_{D/2}^2(0)\right] \frac{\bar{K}^2}{k^2} \right\}. \tag{A.43}$$

Here r is a numerical coefficient which depends on the precise implementation of the regulating function: $r = 0$ for a Type I regulator while the Type II construction of [3] corresponds to $r = 1$. In order to be consistent with the evaluation of the other traces in the gravitational and matter sectors, we will resort to the Type I regulator scheme, setting $r = 0$. Adding the results (A.39), (A.42), and (A.43) to the contribution from the gravitational sector gives rise to the N_S, N_V, and N_D-dependent terms in the beta functions (4.49).

References

1. R. Percacci, D. Perini, Constraints on matter from asymptotic safety. Phys. Rev. D **67**(8), 081503 (2003). https://doi.org/10.1103/PhysRevD.67.081503. eprint: hep-th/0207033 (cit. on p. 167)
2. R. Percacci, D. Perini, Asymptotic safety of gravity coupled to matter. Phys. Rev. D **68**(4), 044018 (2003). https://doi.org/10.1103/PhysRevD.68.044018. eprint: hep-th/0304222 (cit. on p. 167)

Curriculum Vitae et Studiorum

Alessia Benedetta Platania

Personal information
Full name: Alessia Benedetta Platania
Date of birth: 18 October 1990
Place of birth: Catania, Italy
E-mail: alessia.platania@oact.inaf.it

Education and Employments

01-10-09 to 04-10-12	B.Sc. in Physics (Summa cum Laude)-University of Catania
	Supervisor: Prof. Siringo
	Thesis: The Berry geometric phase
01-10-12 to 29-10-14	M.Sc. in Physics (Summa cum Laude)-University of Catania
	Supervisor: Prof. Branchina
	Thesis: Stability of the EW vacuum and Physics beyond the Standard Model
01-11-14 to 23-03-18	Double Ph.D. in Physics (Summa cum Laude)-University of Catania and Radboud University Nijmegen (funded by INFN)
	Supervisors: Dr. Bonanno and Dr. Saueressig
	Thesis: Asymptotically Safe Gravity-from spacetime foliation to cosmology
03-11-17 to present	PostDoc Researcher-Heidelberg University

© Springer Nature Switzerland AG 2018
A. B. Platania, *Asymptotically Safe Gravity*, Springer Theses,
https://doi.org/10.1007/978-3-319-98794-1

Research interests

- Quantum Gravity (Asymptotically Safe Gravity)
- Functional Renormalization Group
- Black holes physics
- Cosmology

Academic awards and fellowships

- *Degree award "Roberto Giordano" 2013/2014* (awarded by the University of Catania, in recognition of her master's degree examination results and thesis)
- *INFN fellowship* (granted by the italian institute for nuclear and particle physics (INFN) for the XXX–cycle Ph.D. Program of the University of Catania)

List of Publications

[1] V. Branchina, E. Messina, A. Platania - *Top mass determination, Higgs inflation, and vacuum stability* - JHEP 09 (2014) 182 [arXiv:1407.4112]

[2] A. Bonanno, A. Platania - *Asymptotically Safe inflation from quadratic gravity* - Phys. Lett. B 750 (2015) 638 [arXiv:1507.03375]

[3] A. Bonanno, A. Platania - *Asymptotically Safe $R + R^2$ gravity* - PoS(CORFU 2015)159

[4] A. Bonanno, B. Koch, A. Platania - *Cosmic Censorship in Quantum Einstein Gravity* - Class. Quantum Grav. 34 (2017) 095012 [arXiv:1610.05299]

[5] J. Biemans, A. Platania, F. Saueressig - *Quantum gravity on foliated spacetimes - Asymptotically safe and sound* - Phys. Rev. D 95 (2017) 086013 [arXiv:1609.04813]

[6] J. Biemans, A. Platania, F. Saueressig - *Renormalization group fixed points of foliated gravity-matter systems* - JHEP 05 (2017) 093 [arXiv:1702.06539]

[7] A. Bonanno, B. Koch, A. Platania - *Asymptotically Safe gravitational collapse: Kuroda-Papapetrou RG-improved model* - PoS(CORFU2016)058

[8] A. Bonanno, G.J. Gionti, A. Platania - *Bouncing and emergent cosmologies from ADM RG flows* - Class. Quantum Grav. 35 (2018) 065004

[9] A. Platania, F. Saueressig - *Functional Renormalization Group flows on Friedman-Lemaitre-Robertson-Walker backgrounds* - Found. Phys. (2018). https://doi.org/10.1007/s10701-018-0181-0 [arXiv:1710.01972]

[10] A. Bonanno, B. Koch, A. Platania - *Gravitational collapse in Quantum Einstein Gravity*. Found. Phys. (2018). https://doi.org/10.1007/s10701-018-0195-7 [arXiv:1710.10845]

[11] A. Bonanno, A. Platania, F. Saueressig - *Cosmological bounds on the field content of asymptotically safe gravity-matter models*. Phys. Lett. B (2018). https://doi.org/10.1016/j.physletb.2018.06.047 [arXiv:1803.02355]

Printed in the United States
By Bookmasters